The Altruism Equation

▲

The Altruism Equation

SEVEN SCIENTISTS SEARCH FOR
THE ORIGINS OF GOODNESS

▲

LEE ALAN DUGATKIN

PRINCETON UNIVERSITY PRESS

PRINCETON AND OXFORD

Copyright © 2006 by Princeton University Press

Published by Princeton University Press, 41 William Street,
Princeton, New Jersey 08540
In the United Kingdom: Princeton University Press,
99 Banbury Road, Oxford
OX2 6JX

First paperback printing, 2022

Cloth ISBN 978-0-691-12590-9
Paperback ISBN 978-0-691-24213-2

Library of Congress Cataloging-in-Publication Data

Dugatkin, Lee Alan, 1962–
The altruism equation : seven scientists search for the origins of
goodness / Lee Alan Dugatkin.
p.   cm.
Includes bibliographical references and index.
ISBN-13: 978-0-691-12590-9 (hardcover : alk. paper)
ISBN-10: 0-691-12590-2 (hardcover : alk. paper)
1 Altruistic behavior in animals   2. Altruism.   I. Title.
QL775.5.D84 2006
591.56—dc22   2006005400

British Library Cataloging-in-Publication Data is available

This book has been composed in Palatino

press.princeton.edu

DEDICATED TO THE MEMORY OF

MY BELOVED FATHER, HARRY DUGATKIN (1918–2004),

WHO TAUGHT ME THAT

FAMILY ALWAYS COMES FIRST

▲

▲ *Contents* ▲

▲ *Preface* ▲

For more than a hundred years, a fierce debate raged within the scientific community on the importance of blood kinship in shaping altruism in everything from animals to humans. This debate, which began in 1859, was extremely contentious, because whoever was right was going to control our view of the origins of goodness. The reason is simple—at its heart, altruism is about incurring a personal cost in order to help others, and that is close to what most of us mean when we speak of doing good. So in essence, a theory on altruism is a theory on goodness.

The debate on altruism and kinship would raise to the surface a number of related questions: was nature red in tooth and claw, or a cooperative paradise? And whatever the answer, was there really a biological theory that could explain this? Before the fighting came to an end, politics, philosophy, mental illness, and even religion would enter the fray, complicating for close to a century attempts to find and settle on scientific answers to scientific questions.

For long stretches of time, the role of blood kinship in shaping human and nonhuman altruism occupied some of the best minds in science. We shall see why four British scientists—Charles Robert Darwin, Thomas Henry Huxley, J.B.S. Haldane, and finally W. D. Hamilton—devoted a good piece of their working lives to the question of kinship and altruism and how that obsession transformed their own lives. Along the way, we will encounter the Russian Prince Petr Kropotkin, the most important anarchist of his time, and two American academics—the Quaker scientist Warder Allee and a suicidal intellectual giant named George Price.

Eventually biology did solve the puzzle of blood kinship and altruism in the form of a mathematical equation developed by a shy, brilliant evolutionary biologist named William D. Hamilton. Hamilton came onto the stage in the 1960s, and using the cost-benefit perspective we most often associate with economists, along with a deep understanding of how evolution operates, crisply and precisely laid out a simple mathematical model explaining why individuals treat blood kin in such a special way.

Phrased in the hard, cold language of natural selection, Hamilton's model boils down to this: blood relatives share many of the same genes, and so by helping kin, you are indirectly helping yourself. Of course, it is a bit more complicated than that, but we will get into the complications soon enough.

Although it took more than a decade for the implications of his work to take hold, Hamilton's model on altruism and blood kinship eventually gained him the ultimate scientific accolade— a rule that bears his name. This rule's influence on evolutionary biology has been as great as the impact of Newton's laws of motion on physics. Yet the story of how it was discovered, how it changed the lives of the men who worked on its discovery, and why even its discoverer wished that the opposing position had turned out to be right, has never before been told.

But first things first. Our story of blood kinship and social behavior begins where all stories of evolution begin—with Charles Darwin.

▲ *Acknowledgments* ▲

THE ACKNOWLEDGMENT section always makes me nervous, as I fear that I may forget to express thanks to one of the many people who helped make this book possible. That said, I am grateful to the following individuals who took time from their very busy schedules to allow me the chance to interview them: Richard Dawkins, Lord Sir Robert May, Alan Grafen, Paul Harvey, E. O. Wilson, Naomi Pierce, David Haig, James Schwartz, Jeremy John, Kern Reeve, Stephen Emlen, Tom Seeley, Paul Sherman, and the late Garret Hardin. I am also indebted to Michael Ruse, Alan Grafen, Paul Ewald, Marc Bekoff, Jeffrey Robbins, and Ryan Earley, who read all or parts of the manuscript and provided essential critiques.

Sam Elworthy, my editor at Princeton University Press, is as good as they come. Sam's patience with a type A, semineurotic author was greatly appreciated, and his input at every level has improved this book dramatically. My agent, Susan Rabiner, has been with me every step of the way, from the book's inception to its completion. As always, Susan's guidance, assistance, and editorial advice were much appreciated.

Finally, I wish to express my sincere appreciation to my wife, Dana, and son, Aaron, who have always stood by my side.

The Altruism Equation

▲

A Special Difficulty
That Might Prove Fatal

Whil e writing *On the Origin of Species* in the late 1850s, Charles Darwin was unencumbered by the strict editorial rules that apply to scientists today. He had the liberty to indulge in wide-ranging digressions that at times became streams of consciousness.[1] This freedom allowed him the scope to tackle issues that he might otherwise have avoided. In particular, Darwin was not afraid to address problems associated with his theory of evolution by natural selection. He did so often, and at length.

This book is about one of Darwin's problems. It began as a small difficulty with honeybees. At first glance, it did not seem like the sort of complication that could sink a theory that many have characterized as the most important one that biology has ever produced. But it turned into a problem that troubled biologists, fascinated naturalists, engaged popular writers and the general public, and even worked its way into political discourse for the next 145 years.

Honeybees had been introduced into Britain around a.d. 45,[2] and by Darwin's day, some five hundred authors had written on bees and beekeeping.[3] By the start of the eighteenth century, England had become the world's leader in the production of apicultural products such as honey and wax, and *The Philosophical Transactions of the Royal Society of London* was an important repository for articles about various aspects of bee life. What's more, the public had fallen in love with bees, particularly when it discovered some of the intriguing natural history of these insects. Bee enthusiasts described how worker bees who were fed "royal jelly" developed into queens and how the same bee egg would develop into a male if it remained unfertilized but become a female if it was fertilized with a drone's sperm.[4]

In practice, what the scientific and public love affair with bees meant was that they could not be ignored in the *Origin*, and as

Darwin biographer Janet Browne notes, Darwin "was specially exorcised over honey bees."[5] If any aspect of bee life was at odds with natural selection, then Darwin understood that it had to be addressed front and center in order for his theory to be credible. One such problem was the existence of nonreproductive—that is, sterile—castes that often occur in insects such as bees, wasps, and ants. These workers are true altruists. In the first place, they do not reproduce but instead provide all sorts of resources to queens—the individuals who *do* reproduce. That alone would make them altruists, in the sense of incurring a personal cost that in turn benefits others. Some, but not all, sterile workers will also defend the hive tirelessly, if need be, with their own lives. This too constitutes an act of altruism, and so the sterile workers who defend the hive are, in a sense, doubly altruistic. And what's more, these bees are designed differently from others in the hive. Differences in size and shape, in fact, allow them to be particularly adept at being altruists.

Sterile social insects were clearly a hurdle for Darwin's theory of natural selection, which posited that only those traits that increased an *individual's* reproductive success would, over subsequent generations, increase in frequency. Sterility and kamikaze-like hive defense would seem to be precisely the sorts of traits that natural selection should operate against, and Darwin knew it.

The process of natural selection, as Darwin saw it, was simple yet extremely powerful: "Natural selection can act only by the preservation and accumulation of infinitesimally small inherited modifications, each profitable to the preserved being." For example, Darwin asked his reader to imagine the wolf that "preys on various animals, securing some by craft, some by strength, and some by fleetness." When prey for wolves are scarce, natural selection acts with brute force on wolf populations. "Under such circumstances," Darwin argued, "the swiftest and the slimmest wolves would have the best chance of surviving and so be preserved or selected. . . . I can see no more reason to doubt this, than that man can improve the fleetness of his greyhounds by careful and methodical selection." Wolves possessing the traits that best suit them for hunting survive longer and produce more offspring—offspring, in turn, who possess the very traits that benefited their parents in the first place. Generation after generation, "slow though the process of selection may be,"[6] noted Darwin, eventually you end up with a wolf better adapted for hunting. There is nothing remotely altruistic going on here: individual wolves do better when they possess certain traits than when they do not, and selection operates to increase the frequency of such traits.

Darwin recognized that natural selection not only operates on morphology (as in the wolf case), but on behavior as well. If behavioral traits were passed from parent to offspring, and these traits had strong, positive effects on longevity and reproductive output, selection would favor such behavioral traits over others. Darwin nicely illustrated how natural selection could operate on behavior by using the egg-laying habits of the cuckoo, a bird notorious for depositing its eggs in the nests of other species. How could such a bizarre trait evolve? What's in it for the cuckoo that such odd behavior should be favored by natural selection?

For Darwin, the potential benefits for parasitic egg-laying behavior abounded. Following his lead, imagine that at the start of this evolutionary process some cuckoos occasionally laid some of their eggs in the nest of another species. Darwin believed that parasitic egg layers might profit "by this occasional habit through being enabled to migrate earlier . . . or if the young were made more vigorous by . . . the mistaken instinct of another species than reared by their own mother." Migrating early and producing more "vigorous" offspring will clearly be favored by the process of natural selection. With such benefits available, if young cuckoos inherited their mother's tendencies to lay eggs in the nests of others, as Darwin thought them "apt" to do, then "the strange instinct of our cuckoo could be, and has been, generated."[7] And again, there is no altruism in play here. As with the wolf case, *if* one variant of a trait—slim, sleek wolf morphology or parasitic egg-laying behavior—is superior to other variants, and if some means exists by which traits are passed from parent to offspring, then natural selection will produce a better-adapted organism.

Evolutionary biologists today recognize that offspring resemble their parents because they inherit their parents' genes. Darwin did not know about genes, nor did he need modern-day genetics for his theory to work. All he needed to realize was that somehow traits that affected reproductive success were passed from parents to offspring. Any Victorian natural-ist worth his salt would have known that offspring resemble their parents, and Darwin was more than a good naturalist, he was a great naturalist.[8]

Since Darwin, of course, Mendel's laws of genetics have be-come a staple of modern biology, and with the current revolu-tion in molecular genetics, we have a deep understanding of how important genes are in shaping virtually every trait. When it comes to genes and behavior, the modern notion that genes are the fundamental unit passed from generation to generation, and hence the target of natural selection, is often referred to as the "selfish gene" approach—a term first coined by Richard Dawkins in his 1976 book, *The Selfish Gene.*[9] For Dawkins, this approach does not imply that genes are selfish in any emotional or moral sense. In fact, he notes, genes are not anything but a se-ries of tiny bits of DNA put together in a particular sequence and orientation, and somehow distinct from other such tiny bits of DNA. Yet genes can be viewed as "selfish," in that the process of natural selection favors those that can somehow or another get the most copies of themselves into the next generation. In many cases, this will simply come down to a gene's coding for a trait that increases the direct reproductive success of the individ-ual in which it resides. But, as we shall see, this is not the only mechanism by which a gene can get more and more copies of it-self into the next generation. There are more indirect, *but equally powerful*, ways for genes to get lots of copies of themselves passed down from one generation to the next.

Natural selection promotes genes that *appear* to be selfish, in the sense of favoring those that maximize the number of copies of themselves that make it to the next generation. Indeed, one of the reasons that Dawkins chose the term "selfish gene" as a metaphor was to emphasize the fact that genes which code for any trait that benefits the species as a whole, or indeed even groups of unrelated individuals, are doomed. Such genes are bound for the evolutionary trash bin because they are not maxi-mizing their chances of being passed to the next generation. Only those genes that are "selfish" make it in the end. Wolf mor-phology and cuckoo behavior fit nicely into the selfish gene framework; altruism and self-sacrificial hive defense in bees do not, or at least so it appears at first glance.

In the case of Darwin's problem with the bees, he was forced to ask how his theory of natural selection could explain the existence of whole castes of insects that never reproduce and yet protect those that do, even at the cost of their own lives. In other words, what's in it for the altruists? Surely such traits should disappear, and fast, if natural selection worked the way it was supposed to. Altruistic worker bees—whom Darwin recognized as undertaking acts that were "profitable" for others in their hive—appeared to fly directly in the face of his logic.

The existence of sterile altruistic castes was an anomaly that had vexed Darwin since the early 1840s. His worries seem to have stemmed, at least in part, from a reading of Reverend William Kirby and William Spence's textbook *Introduction to Entomology*, in which the authors argued that the incredible behaviors of sterile castes were evidence of the divine hand of the Creator in motion.[10] Darwin's annotations in his own copy of Kirby and Spence's book demonstrate his clear frustration with both the authors' ignorance of basic biology—for example, they implied that neuters could breed—and the whole question of sterile castes and what they meant for his own ideas.[11]

Darwin himself had dabbled in small-scale experiments with social insects at Down House, in one case enlisting the help of his children (William, Henrietta, George, Frank, and Leonard) to better understand various aspects of bee behavior, such as their navigational skills from hive to hive.[12] At one point he had "five or six children each close to a buzzing place," at which point Darwin would tell "the one farthest away to shout out 'here is a bee' as soon as one was buzzing around."[13] Then, like a volunteer fire brigade passing buckets of water down a line, the children along the bee's route would continue signaling until the bees reached Darwin. Though this unconventional use of very young researchers helped Darwin understand communication in social insects, these quasi experiments did little to provide an answer to the mystery of the altruistic castes that permeate the social insects.

It is hard to overemphasize just how concerned Darwin was about the problem of sterile animals that helped others through their acts of altruism. That was simply not the way he envisioned natural selection operating, and at times, the problem of the sterile altruists would, as he himself noted, drive him "half mad."[14] So frustrated was he, that in the *Origin*, Darwin summarized the whole topic of sterile castes as "one special difficulty, which at first appeared to me to be insuperable, and actually fatal to the whole theory."[15]

Over the course of many years Darwin tinkered with a number of hypotheses that might reconcile the altruistic caste problem—a problem that centered on insects but had implications for any behavior that involved helping others at a cost to self—with his theory of natural selection. In the end, he speculated on how blood kinship might solve the problem of sterile altruistic insects. A hundred years later these ideas would be formalized through an equation that would be called "Hamilton's rule," an equation that would revolutionize the field of evolution and behavior, but the seeds of which were laid in the *Origin*.

In a section of the *Origin* entitled "Objections to the Theory of Natural Selection as Applied to Instincts: Neuter and Sterile Insects," Darwin proposed that the problem of natural selection's producing sterile individuals that often risk their lives to protect others, and appear designed to do just that, ". . . disappears when it is remembered that selection may be applied *to the family*, as well as the individual, and may thus gain the desired end."[16] Help your blood kin—your family—and you can make up for any costs that you yourself incur. Take the case of the altruistic bees. Even though individual bee altruists often paid a huge cost both by defending the hive and by not reproducing, this cost was made up by the benefits accrued by their family members, and hence altruistic behavior could, in principle, evolve. In addition to acting as hive guards, in his *Species Book*, Darwin hypothesized that selection might favor such sterile workers, as they also specialize on other tasks, such as foraging.[17] This in turn benefits all family members by relieving them of the task of foraging, and eventually it became very clear to Darwin "how useful their production may have been."[18] Blood kinship and interactions among relatives it turned out, was the key to solving Darwin's problems with both sterility and altruism.

Darwin seems to have realized the importance of the role of blood kinship in explaining altruism as early as 1848. In a manuscript dated June of that year, he hinted at its importance in the context of how some hives with sterile castes may "predominate" over other hives, presumably as a result of actions that sterile caste members may undertake to help their kin—in Darwin's words, selection would act on "families and not individuals."[19] Help your relatives and you help yourself, albeit indirectly. These ideas, over the course of the next hundred years, would develop into what is today called "kin selection" theory.

The case Darwin presented amounted to this: natural selection could favor the evolution of sterile castes if individuals in such castes helped their blood kin (which they do), because doing so would help ensure the survival of those individuals that could reproduce—individuals with a hereditary makeup very similar to their own. If kin helped each other, even assuming a large cost of so doing (picture the worker honeybee's suicidal attack on nest predators) the process of natural selection could still favor such a trait, because those being helped were similar in their makeup to those doing the helping. In modern-day terms, genes can increase their frequency in the next generation by aiding the reproduction of copies of themselves that just happen to reside in other individuals—blood relatives. Again, Darwin did not know about genes per se, but he did know that blood relatives resembled one another more than strangers, and this was just enough information to speculate on the role of kinship in the evolution of altruism.

Darwin was still somewhat ambivalent about the power of this explanation in 1848,[20] but over the next decade he became more and more convinced of the utility of his initial explanation—so much so that it found its way into the *Origin*, when so many of Darwin's early arguments did not. One turning point in his thinking on the power of blood kinship in evolution took place when he read William Youatt's work on cattle breeding.[21] As Darwin noted, cattle breeders are interested in producing meat with the "flesh and fat to be well marbled" together. The problem is that to get such meat, a breeder must kill the cattle that produce it. Developing breeding lines of cattle, then, with just the right mixture of flesh and fat marbled together would seem impossible using standard techniques that involve breeding individuals with the desired trait, since in this case, such individuals are slaughtered for their meat. Darwin notes that to solve this problem "the breeder goes with confidence to the same family and has succeeded."[22] In other words, even in the pregenetic era, breeders were aware that blood kin were very likely to resemble one another, and so to achieve a desired trait—in this case, marbled meat—one could breed from blood relatives.

The second intellectual pillar for Darwin's thoughts on kinship and the evolution of altruism derived from his discussions about social insects with his entomologist colleague, Fredrick Smith. As Darwin ultimately saw the situation, "this principle of selection, namely not of the individual which cannot breed, but of the family which produced such individuals, has I believe been followed by nature in regard to the neuters amongst social insects."[23] In some instances, he referred to this as selection at the level of the community (the hive, for example) rather than the kin group per se. Indeed, as Robert Richards details in his book *Darwin and the Emergence of Evolutionary Theories of Mind and Behavior*, Darwin discussed such "community-level" selection in a number of instances.[24] But when it came to altruism and the social insects, the communities to which Darwin referred in the *Origin* were almost always made up of blood relatives. That said, some of Darwin's intellectual descendants would return to community-level selection in their own quests to understand the evolution of altruism.

In one sense, by turning to kinship for the answer, Darwin both posed and solved the conundrum of the evolution of altruism. The problem was confronted, as in the case of sterile insects, and the remedy—what we would now call kin selection—was proposed. But in two very important ways that would haunt the field for a century after the *Origin*, Darwin failed to settle the issue. First, without experiments or some sort of mathematical framework for his theory, he was never able to answer the questions his theory brought forth, namely: precisely how does what we now call kin selection operate? For example, just how does the *degree* of kinship affect the evolution of altruism? Some blood kin, such as parents and offspring or siblings, are very closely related, but other sets of relatives, like second cousins, are much less so. Does that matter, and if so, exactly what does it mean in terms of the predictions one can make regarding the evolution of altruism? Further, does it matter how costly the altruistic act or how large a benefit to the donor of such altruism? If so, how are these costs and benefits to be measured, and what's more, how does ecology affect such costs and benefits?

These fundamental questions required answers. Indeed, in the long run, they would require a mathematical model of kinship and altruism—a model that made specific and testable predictions. Without this, Darwin's ideas on kinship and altruism were akin to a verbal precursor to Einstein's mathematical theory of relativity: nice, but lacking in the hard equations that are needed to establish a bedrock theory. It would take a good hundred years for such models of kin selection theory to appear on the scene.

The second, and in some ways just as important, question that Darwin opened up for debate was one that he never got into, but one that would forever intertwine itself in all future discussions of blood kinship and altruism. And this question—do evolutionary pressures to be kind and generous to others extend beyond blood relatives?—has implications for a much broader audience. From an evolutionary perspective, to what extent do we expect to see generosity as a family affair, and only a family affair?

Consider this: Public outrage follows a judge's order that a child be taken from foster or adoptive parents—maybe the only mother and father the child has ever known—and delivered back to the child's biological parents. But in most such situations, the judge has little choice. Our legal system recognizes blood kinship as a special relationship that society has an obligation to protect, absent some severe aberration in the biological parents that renders them incapable of raising their own child. Of course, just because an idea is codified into law, does not mean that it is scientifically valid. The point here is different: namely, the notion that altruism is particularly pronounced between blood kin is so universally held that it has worked its way into our very legal system. Darwin was silent on such issues. But what started as a scientific matter about social insect evolution ends up having much broader implications.

Those broader implications were understood in Darwin's day as well as our own. For one thing, by directly addressing the problem of honeybee altruism, Darwin not only tackled a major obstacle that stood in the path of his scientific theory, he also further alienated some religious individuals who were already struggling with his idea that natural processes could explain the

9

diversity of life that we see around us. For when it came to altruism, kinship, and social insects, Darwin was not using his theory to suggest how we get a new species of barnacle or earthworm. Rather, he was positing a hypothesis for how self-sacrificial behavior—a subject which, to that moment in time, had been reserved for religion to address—could come into existence. And to make the situation even more complicated, Darwin's ideas clearly meant that altruism could (and did) evolve in creatures other than humans, and that by studying such creatures we could potentially better understand our tendency to be altruistic, particularly toward blood kin.

And not only religious individuals were troubled by what Darwin was saying about kinship and altruism, nor is this surprising. Darwin was on fairly safe ground when discussing complex anatomical structures like the insect eye, because this was not something that the lay public necessarily understands or even cares to understand. But altruism is not like the insect eye. Very few people have their own theories about how the insect eye evolves, but almost everyone has his or her own ideas on why humans are or are not altruistic. These ideas are often spawned from philosophy, religion, and politics, but sometimes arise solely from gut feelings about why we are the way we are. And, of course, scientists too have their philosophical, religious, and political views, and they are not immune from the influence of such ideas on their scientific work; particularly when the questions being studied have, by their very nature, implications for philosophy, religion, and politics.

Over and over we shall see how personal views weave their way into the hundred-year odyssey from Darwin's original ideas to our modern mathematical models of altruism and kinship. Scientists can certainly construct very objective experiments on kinship and altruism, even if they have personal opinions on the subject; it is just more difficult to do so than it is for other topics, because it has such broad implications about the foundations of goodness. And everyone cares about that. Many of the scientists we will encounter seemed almost obsessed with understanding the role that kinship plays in the evolution of altruism—much more so than we see when people study the evolution of almost any other trait. The reason is simple. Unraveling how blood kinship affects altruism would not only be hailed as a major scientific achievement (and it was), but it would tell us so much about our very nature.

Given the central role that the desire to understand goodness—or its absence—holds in human psyche, it is hardly surprising that not long after Darwin published *On the Origin of Species*, the questions surrounding kinship and the evolution of altruism got personal. In 1888, a long-standing argument over kinship and altruism intensified between two of the best-known personalities of their time: Thomas Henry Huxley, "Darwin's bulldog," and Petr Kropotkin, an anarchist former prince of Russia, who would pen a classic book on evolution and kinship entitled *Mutual Aid*. Huxley took his old friend Darwin's ideas to a logical extreme, contending that altruism was rare, but that when it occurred, it was *always* tied to blood kinship. Kropotkin saw things in a radically different way. Altruism (what he called "mutual aid") could be observed everywhere in the world, and Kropotkin was certain it had nothing to with kinship. His fight with Kropotkin would make Huxley better understand Darwin's lament, "I often think my friends (and you far beyond others) have good cause to hate me, for having stirred up so much mud and led them into so odious trouble."[25]

11

Darwin's Bulldog versus
the Prince of Evolution

ENGLAND, 1888: Slowly adapting to life in his home away from home, former Russian prince and well-known anarchist Petr Kropotkin stays vigilant, always keeping an eye out for the Russian secret police, who he rightly believes are out to capture him and place him where they think all anarchists belong—in a dark, dingy jail cell. Having been in and out of Russian and French prisons for his political positions, Kropotkin knows what life in such cells is like—indeed, just a year earlier, he had written a whole book on that subject.[1] To keep his sanity and avoid paranoia, he spends his time reading and writing.

Kropotkin's reading material often includes the popular British journal, the *Nineteenth Century*, home to such writers as Alfred Lord Tennyson, Beatrix Potter, Baron Rothschild, and Prime Minister William Gladstone. As he scans the table of contents of the February 1888 issue, Kropotkin's political instincts immediately hone in on an article by Thomas Henry Huxley entitled "The Struggle for Existence." As Kropotkin peruses this article—an essay that mixes politics and science—one paragraph catches his attention. Here, Darwin's close confidant, Professor Huxley, notes in no uncertain terms:

> From the point of view of the moralist, the animal world is on about the same level as a gladiator's show. The creatures are fairly well treated, and set to fight; whereby the strongest, the swiftest and the cunningest live to fight another day. The spectator has no need to turn his thumb's down, as no quarter is given.... The weakest and the stupidest went to the wall, while the toughest and the shrewdest, those who were best fitted to cope with their circumstances, but not the best in any other sense, survived. Life was a continual free fight, and beyond the limited and temporary relations of the family, the Hobbesian war of each against all was the normal state of existence.[2]

Huxley's point is clear: altruism is hardly ever to be expected, and when it is, it will always be tied to blood kinship. Kropotkin is livid, beside himself with anger over what he calls Huxley's "atrocious article." His own five-year expedition through Siberia has shown him a world precisely the opposite of the one Huxley describes. He sees altruism at every turn in nature, and what's more, as he understands it, such behaviors have nothing to do with family relations and blood kinship. Indeed, in short order, Kropotkin is penning articles and books, pleading with his readers, "Don't compete!—competition is always injurious to the species, and you have plenty of resources to avoid it. . . . That is the watchword which comes to us from the bush, the forest, the river, the ocean. Therefore combine—practise mutual aid! That is what Nature teaches us; and that is what all those animals which have attained the highest position in their respective classes have done."[3]

So begins the first battle in a century-long war on the question of altruism and cooperation and their ties with blood kinship. Although this war would eventually be resolved via equations and empirical evidence, it started with Huxley and Kropotkin—two people with diametrically opposed political and philosophical worldviews. In a sense these two men represented the personification of so many of the battles being played out in the late nineteenth century—East versus West, presocialist anarchism versus Victorian competition, nature as the pinnacle of cooperation versus nature "red in tooth and claw." Kropotkin and Huxley's arguments, ostensibly over cooperation and blood kinship, in fact involved fifty thousand–mile treks into Siberia and the Amazon rain forest, anarchy, socialism, mental disease, religion, and politics, all weaving their way into the very heart of science. Each would claim that nature was behind him: Huxley in his assertion that nature was a bloodbath, with the family as the sole repository of goodness, and Kropotkin in his adamant defense of nature as a bastion of altruism, where blood kinship plays almost no role. What's more, they both approached this question with armloads of baggage. Who Kropotkin and Huxley were, and where they came from, had profound implications for what they believed to be true about human and animal altruism, and its tie to kinship; and it explains why they vehemently disagreed about these questions.

Thomas Henry Huxley was not born into one of the first families of England, but his reputation later raised his family to that status, and he bequeathed that legacy to his descendants, including, most notably his grandsons, Aldous and Julian.[4] Born on May 4, 1825, in Ealing, Middlesex, Thomas was the last of six children. His father, George Huxley, was one of the masters at an Ealing school. Of all his qualities, Thomas attributed only his skill at drawing and his "hot temper" to his father. In direct contrast, Thomas's mother, Rachel, was a Cockney ball of energy and the love of young Huxley's life. He adored her and credited all the admirable characteristics he possessed, "physically and mentally," to his mother. He reminded people that he was the son of his mother "completely." Most important for Thomas, Rachel possessed a "rapidity of thought." She, however, took little credit for this talent, shrugging it off with, "I cannot help it, things flash across me."[5] In any case, Thomas had his mother's quick wit and intuition.

Huxley's England was at that time going through a massive economic depression that, according to his biographer Adrian Desmond, was characterized by "tens crammed into a room, babies diseased from erupting cesspits and the uncoffined dead gnawed by rats."[6] It was a country where religion and politics had utterly and completely failed to stop the worst economic catastrophe in a hundred years. The society in which Huxley was reared seemed a classic case of what would soon be called "survival of the fittest," with little room for altruism of any sort. In his short autobiography, he describes his childhood days in no uncertain terms, and with the analytical skills that would be evident in his later scientific writings: "We boys were average lads, with much the same inherent capacity for good and evil as others; but the people who were set over us cared about as much for intellectual and moral welfare as if they were baby-farmers. We were left to the operation of the struggle for existence among ourselves."[7] Altruism and cooperation were nowhere to be found in that world, except at home with his mother and siblings—that is, with blood kin.

Thomas often retreated from the tough reality in which he lived, one that he "hated and avoided," into a fantasy world, a world of philosophy and science. Even at age twelve, the young Huxley "was not satisfied with the ordinary length of the day,"[8] and extended the time available for learning by nighttime, candlelit readings of such books as Hutton's revolutionary *Theory of the Earth*. His strongest link to the world surrounding him was his connection with the men of natural science, minds that he actively sought out. To Thomas, it was the scientists, not the clergymen, who would be England's saviors.

For a young man with Thomas's interests in philosophy and science, the normal seven-year apprenticeship in some craft would never do. Although he never particularly cared "about medicine as the art of healing," Huxley was nonetheless interested in this field, insomuch as it could provide him something he was keen to possess—an understanding of "the mechanical engineering of living machines."[9] Luckily, some of his in-laws were medical doctors. One of them, James Charles Cooke, was a particularly flamboyant gentleman, described by Adrian Desmond as a "beer-swilling, opium-chewing man of massive medical lore."[10] Cooke took on thirteen-year-old Thomas as a medical apprentice, and in the process, demonstrated to Huxley that family will often go to great lengths to help one another.

Under Cooke, Thomas's first experience in the world of medicine was emotionally devastating—a face-to-face encounter with a naked cadaver in a dreary dissecting room. Here, for Huxley was hands-on experience with what Darwin would soon call the struggle for existence. Then, in January 1841, after serving with Cooke, Huxley moved on to work as an assistant to one Thomas Chandler, a "lowlife doctor in the east end,"[11] where he worked in a small "clinic" located in the Rotherhithe section of London. The world he encountered there was more macabre than any he had experienced in his short life so far, one of horrifically unsanitary conditions and starving children dying of disease. The forces that shaped Huxley's psyche, and that shaped the human condition as he would see it, were beginning to take root. Nature, it appeared, could be brutal and showed no inherent tendencies whatsoever toward altruism. If Thomas wanted to see altruism, he turned to his own kin.

Medical school, as well as his earlier apprenticeships in which people came to him for medical aid but "were really suffering from nothing but slow starvation,"[12] gave Huxley practical experience with the filth and disease that so infested the England of that day—yet another encounter with nature as a bloodbath, devoid of altruism and cooperation, with no retreat but the family unit. After he finished medical school in the spring of 1846, in order to cover "a mounting pile of debts,"[13] Huxley began a four-year stint as a surgeon and scientist on the HMS *Rattlesnake*; a journey that we shall revisit later. Upon his return in 1850, Huxley began a slow but steady rise up the academic and scientific ladder. When courting his fiancée, Henrietta Heathorn, in July 1853, he prophesized, "I will make myself a name and a position as well as an income by some kind of pursuit connected with science, which is the thing for which Nature has fitted me best, if she has fitted anyone for anything."[14] And so it was to be, for by the time *On the Origin of Species* was published, less than a decade after the *Rattlesnake* returned to port, Huxley was already recognized as one of the leading intellects in British science.

As one of the preeminent British scientists of the nineteenth century, Thomas Henry Huxley was no stranger to controversies surrounding evolution: one of his admitted occupations was to engage in "an endless series of battles and skirmishes over evolution."[15] As a case in point, on the day before the *Origin* was published, in an attempt to calm Darwin, Huxley wrote to his colleague, "As to the curs which will bark and yelp, you must recollect that some of your friends, at any rate, are endowed with an amount of combativeness which (although you have often and justly rebuked it) may stand you in good stead. I am sharpening up my claws and beak in readiness . . . ," and he added, for good measure, "prepared to go to the stake, if requisite."[16]

Huxley could back this bravado with action, as he knew Darwin's ideas as well as anyone, and he had the public-speaking skills to promulgate them. Furthermore, he was a talented and prolific writer who held back no punches—indeed, the *Pall Mall Gazette* spoke of Huxley's pen as a formidable weapon: "Cutting up monkeys was his forte, and cutting up men was his foible."[17] Darwin referred to him as "my good and kind agent for the propagation of the Gospel,"[18] while others simply used the moniker that Huxley created for himself—"Darwin's bulldog."

Once Darwin developed his theory of natural selection, it was Huxley who spread the word for him. Darwin was suffering from intense gastrointestinal problems, as well as exhaustion and dizziness, which made him feel as though he was "confined to a living grave," and was often too ill to speak in public about evolution. And so he relied on Huxley to present and defend his ideas to the people, the media, and other scientists; and Huxley was happy to do so, the only other option in his mind being "to let the devil have his own way."[19]

It was in the midst of "reading the *Origin* slowly again for the nth time,"[20] to compose a much belated Royal Society of London obituary for Charles Darwin, that Huxley wrote his famous (or infamous, depending on one's perspective) 1888 "gladiator essay"—the article that laid out the ideas on altruism and kinship that so infuriated Prince Kropotkin. For Huxley, this piece was, more than anything else, a tribute to his friend Darwin. In his position as Darwin's prolocutor, and from his understanding of natural selection and the critical role that competition and death played in this process, Huxley felt he had no choice but to accept and promulgate the idea that nature was a bloodbath; hence the quote from his gladiator essay that so angered Petr Kropotkin. For Huxley, there was but one respite from nature's carnage, and that lay in "the relations of the family"—in other words, in blood kinship. As Huxley saw things, it was only among family members that the dog-eat-dog rules that apply elsewhere in nature were relaxed and the observer might see acts of goodness. Such goodness was never to be taken for granted, even among blood kin; but if it was to be found anywhere, that is where one should be looking. The beauty of Huxley's view is how straightforward it is; the reader is left with no sense of ambiguity when it comes to Professor Huxley's version of evolution, altruism, and blood kinship.

The gladiator reference in his 1888 essay was not a metaphor for Huxley; he meant it literally, and he was quite clear about the implications of his view of nature. In his now classic speech/essay, "Evolution and Ethics," Huxley adamantly warned his readers to "understand, once and for all, that the ethical progress of society depends, not on imitating the cosmic process [evolution], still less in running away from it, but in combating it."[21] For Huxley, carnage was the result of the struggle for existence—the family unit being the one place where the occasional altruistic act might be expected. This was no model for human society, and Huxley vehemently separated our "wild zo-ological nature from our ethical existence."[22]

Nature was neither moral nor immoral for Huxley, but rather amoral, demanding ". . . nothing but a fair field and free play for her darling the strongest." In the natural world, no moral judg-ment is to be made between the victors and the vanquished. The deer killed by the wolf is no more noble than the wolf that slaughters it; both were designed to be what they are, and cer-tainly neither was altruistic. Huxley did not turn to Mother Nature for answers about human morality. He put the onus on *Homo sapiens* themselves. Huxley viewed man's evolutionary past as nothing to be proud or ashamed of, for *Homo sapiens*, like all species, simply "floundered amid the general stream of evo-lution, keeping its head above water as it best might, and think-ing neither of whence nor whither." The reason was simple: hu-mankind, like all other species, if left to its predispositions, was caught in an evolutionary trap, multiplying more quickly than its resources. In Huxley's England alone, "about every hundred seconds, or so, a new claimant to a share in the common stock or maintenance presents him or herself among us." This inevitably leads to competition and the struggle for existence; once these are in play, it remains just a matter of time before the gladiator show begins. The natural world held little space for altruism. But, Huxley believed, if man tried "to escape from his place in the animal kingdom," there was always hope that we might "found a kingdom of Man, governed upon the principle of moral evolution."[23] Without such an escape, for Professor Huxley, altruism would remain a family affair, and only a family affair.

Looking back at Huxley's position on altruism and its relation to blood kinship, it is difficult to assess whether the evidence of the day supported his contention or not; he presented the reader with no such evidence to judge, except for emphasizing the well-known fact that parents—even in the animal world—were generally altruistic toward their offspring, providing them with resources at a cost to themselves. Instead, Huxley seems to have reached his conclusions about altruism and kinship based on how natural selection *had* to operate, and equally, if not more im-portantly, on both his life experiences and on the work of one his favorite philosophers, Thomas Malthus.

No single event in his personal life affected Huxley's view of altruism, kinship, and nature more than the death of his daughter Mady.[24] In November 1887, the year before he published his gladiator essay, Thomas Henry Huxley's beautiful and talented twenty-eight-year-old artist daughter, Mady, died of complications related to a mental illness. Mady, whose art had taken numerous prizes, was the apple of her father's eye, and the family did all they could to fight her illness in a time when it was hardly recognized as a disease in the first place.[25] Though a technical diagnosis would have been beyond the medical knowledge of the time, Mady's extreme mood fluctuations, in conjunction with her great artistic talent, suggest bipolar disorder (manic depression). Despite the attention of Jean-Martin Charcot, a leading neurologist specializing in "hysteria," Mady was gone.

Huxley decried what he saw as nature's massacre of the innocents. In his despair over Mady's passing he wrote, "you see a meadow rich in flower and foliage and your memory rests upon it as an image of peaceful beauty. It is a delusion. . . . Not a bird twitters but is either slayer or slain . . . [and] not a moment passes in that a holocaust, in every hedge & every copse battle murder and sudden death are the order of the day."[26] It was in this light, the light of nature as the embodiment of struggle and destruction, nature as the antithesis of altruism (kinship-based or not), that Huxley saw the death of his beloved daughter. At the time, he himself was in the midst of a bout of severe depression, and he described himself as "melancholy as a pelican in the wilderness."[27] Such was his mood when he composed the 1888 essay that Petr Kropotkin so despised.

If Mady's death was the defining moment in terms of how Huxley's personal life affected his views on nature, altruism, and kinship, then his incorporation of Thomas Malthus's views on population growth represents the turning point in the intellectual and philosophical development of his thoughts on altruism and kinship. In 1798, the Reverend Thomas Malthus published a short but extremely influential pamphlet entitled *An Essay on the Principle of Population, as It Affects the Future Improvement of Society*. This work served as a critique of the ideas of Enlightenment philosophers like Condorcet who thought that human society was malleable and infinitely perfectible. Malthus believed that such a view was nothing short of tripe, as its premise went against his "natural law," namely that: "Population, when unchecked, increases in a geometric ratio. Subsistence increases only in an arithmetic ratio." This necessarily created "a strong and constantly operating check on population."[28]

For Malthus, the principle of population—his natural law—led to "misery and vice" in humans. This was the inevitable consequence of limited resources for populations that grow quickly. His view necessitated a pessimistic attitude toward human societies: "I see no way by which man can escape from the weight of this law which pervades all animated nature," he noted. Furthermore, "No fancied equality . . . could remove the pressure of it even for a single century. And it appears therefore to be decisive against the possible existence of a society, all members of which should live in ease and happiness, and comparative leisure."[29] It is difficult to imagine how any altruism—kin-based or not—is possible in such societies.

Malthusian doctrine quickly became widely accepted in England and was eagerly picked up by Charles Darwin. Indeed, Malthus's ideas on population growth became an integral part of Darwin's theory of evolution by natural selection. Darwin recalls his first reading of Malthus as a seminal event in the early formation of his ideas: "In October 1838, that is, fifteen months after I had begun my systematic enquiry, I happened to read for amusement Malthus on *Population*, and being well prepared to appreciate the struggle for existence . . . it at once struck me that under these circumstances favourable variations would tend to be preserved, and unfavourable ones to be destroyed. The result of this would be the formation of new species. Here then I had at last got a theory by which to work."[30] And work it he did, for the competition generated by overpopulation became one fixture of Darwin's ideas on natural selection and the struggle for existence. Darwin went as far as describing his ideas on natural selection as ". . . the doctrine of Malthus applied with manifold force to the whole animal and vegetable kingdoms."[31]

These ideas were ingrained not only in Darwin's views, but in those of Huxley as well. In 1873, a good fifteen years before the gladiator essay, Huxley left no room for doubt about his view of the relationship between Malthus and natural selection: "It [natural selection] is indeed simply the law of Malthus exemplified . . . ," Huxley noted, and "although he was much abused for his conclusions at the time, they have never been disproved and never will be—he [Malthus] showed that in consequence of the increase in the number of organic beings in a geometric ratio, while the means of existence cannot be made to increase in the same ratio, that there must come a time when the number of organic beings will be in excess of the power of production of nutriment, and thus some check must arise to the further increase of those organic beings."[32] Darwin was able to generate natural selection from Malthusian ideas, but Huxley was unable to generate altruism.

Huxley was an ardent Malthusian who believed that our "deep-seated organic impulse"[33] toward rapid multiplication led to a Hobbesian war of all against all when resources were scarce. His conviction on the validity of this process turned him away from evolution for any guidance on morality. And just as important, they set the stage for his seeing nature as a bloodbath in which the only decency that could be expected was that between family members. What else could nature be if Malthus was right?

Petr Kropotkin thought otherwise. For Kropotkin, and what has been called the Russian school of evolutionary biology, Thomas Malthus was nothing short of a villain who had been co-opted to promote the misguided view of evolution that people like Huxley championed. While overpopulation and the heartless competition and dearth of altruism that it produced may have been part and parcel of the world of the English, it was foreign, in every sense of the word, to Russians like Kropotkin. To see why, we need to put things into geographic perspective. When Darwin published the *Origin*, Russia made up one-sixth of the earth's dry land mass, with Siberia alone being forty times larger than Great Britain and Ireland combined.[34] Yet this vast expanse was inhabited by a mere 82 million people (as compared with the 35 million inhabitants of the British Isles), in part owing to the very harsh weather, in which vast sections of the country would "stay frozen eight months out of ten . . . while the rivers freeze all the way to the Black Sea."[35] In such a world, underpopulation, not overpopulation, was the most pressing problem. And Darwin's direct competition did not stem from underpopulation;

hence, instead of evolution via overpopulation leading to nature's cycle of slaughter as per Malthus and then Huxley, underpopulation opened the door to altruism and cooperation for Russian scientists like Kropotkin. And underpopulation allowed the Russians to take the evolutionary processes proposed by Darwin and derive altruism from them.

Initially, Darwin's work on evolution was well received in Russia. A. O. Kovalevskii, a Russian embryologist, recalls that "Darwin's theory was received in Russia with profound sympathy. . . . It immediately received the status of full citizenship and ever since has enjoyed widespread popularity."[36] And indeed, Darwin's ideas *on the whole* were generally accepted in the Russian community, except for those tied directly to Malthus, whom many Russians eventually came to see as little more than a "pastor thief."[37] When *On the Origin of Species* was published in 1859, most Russian scientists did not even know about Malthus. His original essay did not migrate to Russia until 1818—twenty years after it was originally published—and then only in English. The first Russian translation appeared nine full years after the *Origin* was in print.[38] When the scientists and the politicians of that day encountered Malthus's ideas, with a few exceptions, they uniformly rejected his "natural law." The Russia in which they lived bore no resemblance to this Englishman's overpopulated world, and it was not long before his ideas were rejected by Russian economists.[39] For Russians, although Malthus's ideas may have had some application in England, they did nothing to explain life in Russia. Kropotkin himself railed against "the supposed pressure of population on the means of subsistence," which he saw as ". . . mere fallacy, repeated, like many fallacies, without even taking the trouble of submitting it to a moment's criticism."[40]

But for Kropotkin and most of the Russian intellectual community, it was not so much the difference between Malthus's view of the world and theirs that most disturbed them, as their impression that his work reeked of British individualism. One of those Russian intellectuals, N. Danilevskii, summed it up thus: "The essential, dominating characteristic of the English national character is love of independence, the all-sided development of the personality, and individualism; which manifests itself in a struggle against all obstacles presented by external nature and other people. Struggle, free competition, is the life of the Englishman; he accepts it with all its consequences, demands it as his right, tolerates no limits upon it."[41] Clearly, this was not the feudal world in which Russian scientists and laymen found themselves.

Kropotkin and his Russian colleagues, and Huxley and his English colleagues, differed sharply over how to best interpret the ideas of Darwin and Malthus, particularly in terms of what these theories meant for altruism and kinship. The Russians faced a dilemma. They thought highly of Darwin and his work on evolution. Indeed, Darwin's theory of evolution per se received a smoother reception in Russia than in England or the United States. At the same time, for most Russians, Darwin's tie to Malthus was unacceptable, thus creating their quandary: how to accept the former, but reject the latter?[42]

Those outside the scientific community simply dismissed both Darwin and Malthus. Karl Marx chided Darwin as a man "who recognizes among beasts and plants the society of England,"[43] a Malthusian society extraordinaire. Tolstoy, too, weighed in on Malthus and Darwin, and his pen was equally harsh on both of them.[44] He was merciless when it came to Malthus, whom he considered "a very poor English writer, whose works are all forgotten, and recognized as the most insignificant of the insignificant." As for Malthus's natural law, it was "fictitious" and the height of "frivolity" and "stupidity."[45] When it came to Darwin, Tolstoy was no more sympathetic, if slightly less vitriolic. In one fell swoop, he damned Darwin and his ideas, as they "will not explain to you the meaning of your life nor will they provide guidance in your actions"[46]—especially, Tolstoy might have added, with respect to altruism.

Within the Russian scientific community that was home to Kropotkin, the solution to the problem of Darwin's tie to Malthus was to admit that the link existed, but to claim that it was far overblown, that Darwin's ideas worked perfectly well without the Malthusian component. In addition, when they separated Malthus from Darwin, the Russian school discovered what they called "mutual aid," that is, altruism and cooperation—and lots of it.[47] And all that altruism and cooperation had little, if anything, to do with family relations and blood kinship.

For Russian scientists like Kropotkin, who wanted to salvage Darwin and altruism while scrapping Malthus, the fundamental issue on which they focused was the nature of the "struggle" that exists in nature. For this group, Darwinism had been hijacked and inexorably linked with the sort of struggle Darwin imagined between two hungry canids—a contest between individuals that inevitably led to the carnage Huxley described in the gladiator essay. Instead, what was truly important to the "mutual aid" camp of Kropotkin and his colleagues was the sort of metaphorical struggle Darwin described when he wrote that "a plant on the edge of the desert is said to struggle for life against drought."[48] That struggle, particularly when it occurred in animals, would lead to altruism and cooperation, as animals would unite to fight their harsh environment and, in the process, help one another, regardless of blood kinship. The Russian camp was infuriated that so many people had confused these very different forms of competition and their implications for understanding altruism.[49] Kropotkin argued that the "phony Darwinists"[50] were omnipresent, and that such "vulgarizers of the teachings of Darwin have succeeded in persuading men that the last word of science was a pitiless individual struggle for life."[51] At the top of Kropotkin's list of such vulgarizers stood Thomas Henry Huxley.

But there was more to Kropotkin's distaste for Huxley than that. An integral part of his disagreement stemmed from the fact that Kropotkin grew up in a Russia that was in the midst of a political experiment with anarchy and socialism. And Kropotkin was not merely raised in such a world, he grew to be one of its revolutionary leaders—one whose views on both politics and altruism/blood kinship began to congeal in his teenage years and were permanently welded together shortly thereafter, when he took an incredible five-year naturalist's journey through Siberia.

Petr Alexeivich Kropotkin was born in the well-to-do Old Esquerries section of Moscow on December 21, 1842, during the reign of Czar Nicholas I. The Kropotkins had once ruled as grand princes in the Smolensk region of Russia. They claimed descent from the Rurik family, who ruled Russia before the Romanovs, and Petr's father took great pains to make sure that everyone knew that bit of information. Although the family had lost its strongest hold on power some time before Petr's birth, they still lived the life of country gentlemen, with some twelve hundred serfs to work their land and tend to the manor. Russia at the time of Petr's birth was a world where feudalism was entrenched, and the Kropotkins were considered aristocrats, enjoying all the amenities that go along with such stature. His childhood was filled with elaborate balls and parties, and the Kropotkins, though not as rich as others in their class, spared nothing when it came to conspicuous consumption; as Petr recalled it, on any given day they might have "four coachmen to attend a dozen horses, three cooks for the masters and two more the servants, a dozen men to wait upon us at dinner-time . . . and girls innumerable in the maidservant's room."[52]

As a young boy, Petr and his brother Alexander were taught French, German, and Russian by tutors. One tutor in particular, a man named Poulain, would regale Petr with his war stories and was the first to introduce him to democratic ideas. He was a liberating influence on Petr and took pride in telling him of the renunciation of all aristocratic titles by French rebels such as Honre-Gabriel Riguetti—the former count of Mirabeau—a story that entranced the young prince.[53] Poulain so inspired Petr's imagination and his liberal sense of justice, that after finishing his studies under him, Petr referred to himself from that point forward simply as Petr Kropotkin, permanently omitting the title "prince."

Tutors aside, Petr's father (also named Alexander) had one goal in mind for his sons from their birth: a military career. Despite Petr's reservations about such a path (he leaned toward a career in teaching)[54] at the age of fifteen, he was sent to St. Petersberg to enter the Corps of Pages, the most select military school in Russia, which produced the czar's personal pages.[55] During the last days of his schooling in the corps, two critical events in his life converged. First, at the prompting of his brother, Kropotkin read Darwin's *On the Origin of Species*, and it is at this point that he developed his lifelong love of science, particularly evolutionary biology and geology. Even at this early stage of his development as a scientist, Petr seems to have been immersed in these topics; he and his brother had a long, detailed correspondence that "lasted many years" over such issues as the variability of species and how traits are transmitted through time.[56] Kropotkin had not developed his ideas on mutual aid or its relation to kinship at this point, but he was clearly a confirmed evolutionist.

The second event occurred in June 1861. Petr, as the top student in the entire corps, was made Sergeant of the Corps of Pages. Among other benefits, the sergeant was automatically given the job of personal page to the czar. At the start of his service to Alexander II, the liberator of the serfs, Petr had held the czar "in great admiration," even musing on how he would risk his own life and limb to protect him.[57] But as he saw the misery that was truly in store for the freed serfs and others in the underclass, Kropotkin's admiration began to wane and doubts about Alexander II emerged. Kropotkin had some personal liking for the czar, but his attitude toward the state was another story. He was starting to believe that state government was the problem, not the solution, to society's troubles and that centralized authority made mutual aid—altruism and cooperation—impossible. This marked the birth of a philosophy of anarchism that would permeate not only Kropotkin's politics, but his thoughts on altruism and its ties—or more precisely in Kropotkin's eyes, its lack of a tie—to blood kinship.

Ironically, Kropotkin's position in the Corps of Pages enabled him to escape from the world of the royal court, for in March 1862, when it came time for each page to choose his military assignment, head student Petr had first call.[58] By this point in his development, Petr knew that he would not spend his life in the military. He used his rank to choose a post that would allow him to pursue his interests in evolution, natural history, and politics, while maintaining only a nominal military component. His thoughts had "turned more and more to Siberia," in particular to the Amur region that had recently been annexed by Russia. An appointment to such an unexplored region would give him the chance to study and describe a brave new world as his hero, Alexander Humboldt, had done. And so, on June 24, 1862, Kropotkin began what turned into a fifty thousand–mile trek through Siberia—a much colder version of Darwin's voyage on the *Beagle*, which would spur on Kropotkin's passion for studying evolution and mutual aid.[59]

In Siberia, Kropotkin observed what appeared to be altruism and cooperation among both animals and peasants of the region at every turn. Animals united to protect themselves against the harsh realities of life in the far north and in the process helped one another—all, as far as he could tell, in the absence of blood kinship of any sort. For Kropotkin, animals formed herds to find food and keep themselves and others in their herd warm. Peasants lived in small, self-sufficient groups, again fighting the environment by aiding one another in all manner of things. "In all the scenes of animal lives which passed before my eyes," he noted, "I saw mutual aid and mutual support carried on to an extent which made me suspect in it a feature of the greatest importance for the maintenance of life, the preservation of each species and its further evolution."[60] Here lay the seeds of his ideas on evolution, behavior, and politics.

Kropotkin developed his ideas on mutual aid, kinship, and evolution under rather difficult circumstances. His numerous journeys through Siberia—five major expeditions in all—were, in fact, the stuff from which movies are made.[61] He often rode hundreds of miles at a time on horseback or sled through the icy tundra, where temperatures dropped to sixty degrees below zero, Fahrenheit. He traversed unforgiving mountains with nothing to sustain him but bread and tea. At times he would travel ten thousand miles of tundra in a few weeks, even if that meant that he would occasionally have to assume an alias (Petr Alexiev), as he did when crossing into Chinese Manchuria.

Kropotkin seemed immune to the harsh Siberian environment: "The snow covered roads are excellent," he wrote, "and although the cold is intense, one can stand it well enough. Lying full length in the sled . . . wrapped in fur blankets, fur inside and fur outside, one does not suffer much from the cold, even when the temperature is forty or sixty degrees below zero, Fahrenheit."[62] Siberia may have been a frozen wasteland for much of the year, but to Kropotkin it brimmed over with animals that were helping one another, in part as a result of the extreme natural conditions they faced. And, as far as he could gauge the situation, all of this mutual aid was in the complete absence of kinship.

Mutual aid saturated not only the ground of Kropotkin's Siberia, but the air and water as well:

> . . . wherever I saw animal life in abundance, as for instance, on the lakes . . . ; in the colonies of rodents; in the migrations of birds . . . and especially in a migration of fallow-deer which I

witnessed on the Amur, and during which scores of thousands of animals came together from an immense territory . . . in all these scenes of animal life which passed by my eyes, I saw mutual aid and mutual support carried on to an extent which made me suspect in it a feature of the greatest importance for the maintenance of life, the preservation of each species, and its further evolution.[63]

And, again, as far as Kropotkin could tell, almost none of this cooperation and altruism was linked to blood kinship and family life.

Many of the cases of mutual aid Kropotkin wrote about were presented in the laundry-list format above. He would often roll off a long series of examples of what he saw as mutual aid in a seemingly endless, rambling paragraph. It is not uncommon to read in a single paragraph of his book, *Mutual Aid*, of a herd of horses forming a defensive ring to ward off a predator, a pack of wolves hunting as a united front, a beautifully coordinated migrating group of deer, and a bunch of kittens playing with each other.[64]

Kropotkin was an excellent naturalist, and many of his Siberian observations have been confirmed repeatedly. His interpretation of what he saw is another issue. Today, much of what he called mutual aid would not be considered cooperative or altruistic by those studying evolution and behavior. Cooperative and altruistic acts are typically defined in modern-day evolutionary biology as behaviors that benefit others but entail a cost to the individual performing them.[65] For Kropotkin, by contrast, group life per se, and indeed almost every sort of action involving members of the same species—with the possible exception of aggression, which he hardy ever recorded—constituted altruism.

Some of the mutual aid that Kropotkin documented, however, would indeed be classified as altruistic even today. Take the case of his observations of burying beetles, who "must have some decaying organic material to lay their eggs in, and thus to provide their larvae with food; but that matter must not decay very rapidly." Kropotkin wrote admiringly of the burying beetle solution: "So they are wont to bury in the ground the corpses of all kinds of small animals which they occasionally find in their rambles. As a rule, they live an isolated life, but when one of them has discovered the corpse of a mouse or bird, which it could hardly manage to bury itself, it calls four, five, six or ten other beetles to perform the operation with united efforts; if necessary, they transport the corpse to a suitable soft ground; and they bury it in a very considerate way, without quarrelling as to which of them will enjoy the privilege of laying its eggs in the buried corpse."[66] This sort of action entails a cost to the actor and benefit to others, and so meets the modern definition of cooperation. But as always, Kropotkin saw no link between this behavior and blood kinship—no Huxley-like gladiator show was needed to understand the actions of such cooperative insects.

And then there was the mutual aid among humans in the desolate Siberian landscape. "The constructive work of the unknown masses," Kropotkin wrote, "which so seldom finds any mention in books, and the importance of that constructive work in the growth of forms of society, fully appeared before my eyes in a clear light . . . to see the immense advantages they got from their semi-communistic brotherly organization . . . to live with the natives, to see at work the complex forms of social organization which they had elaborated far away from the influence of any civilization, was, as it were, to store up floods of light which illuminated my subsequent readings."[67] Many year later, when Kropotkin's ideas on mutual aid and its independence from blood kinship had gelled in *Mutual Aid*, he supplemented his Siberian experience on human altruism with other cases, consecutively examining mutual aid in "savages," in "barbarians," in medieval guilds that were formed to provide "a close union for mutual aid and support, for consumption and production, and for social life altogether," and in modern man, and noting how this is just what we should expect, and that anything else would have been "quite contrary to all we know of nature."[68]

Kropotkin began his discussion of mutual aid in humans by dismissing an argument posed by Thomas Hobbes, to whom Huxley had referred in support of his gladiator depiction of a nature lacking in altruism, except occasionally among kin. Hobbes, in addition to arguing that nature was a "war of all against all," had speculated that man's original social structure was the immediate family group. This, of course, fits in very nicely with Huxley's subsequent argument that blood relations were the one repository for goodness. But for Kropotkin, this notion was dangerous nonsense, and he was certain that the facts were on his side, as "science has established beyond any doubt that mankind did not begin its life in the shape of small isolated families." Instead, he believed that the initial bed of human altruism was not the family, but the tribe or group, in direct opposition to "Hobbesian speculations," not to mention Huxleyian speculations.[69]

Today, evolutionary biologists almost uniformly agree that Kropotkin was wrong, and that the family most likely did serve as the origin of subsequent human groupings, but it is fascinating to examine the sort of evidence Kropotkin mustered for his view. For example, in attempting to demonstrate that mutual aid was an integral feature of life as far back as the "glacial period," the prince used indirect archeological evidence that he stretched for all it is worth. Kropotkin noted that "isolated finds of stone implements, even from the old stone age, are very rare; on the contrary, wherever one flint instrument is discovered others are sure to be found, in most cases in very large quantities." In his mind, the implication of such information was clear. If archeologists consistently uncover large quantities of stone implements in a given place, then humans must have been living in at least moderate-size groups—groups that were too large to be families—and that such groups must have practiced mutual aid. Even "at a time when men were dwelling in caves, or under occasionally protruding rocks," Kropotkin claimed, "they already knew the advantages of life in societies."[70] Not all of Kropotkin's arguments about the role of mutual aid in human evolution rested so precipitously on extrapolations from indirect archeological evidence, but many did.

Today, work from biology, anthropology, and ethnography clearly shows that in human hunter-gatherer societies, groups and villages are often composed of extended family members, and so fossil remains indicating life in small to moderate groups would be consistent with societies composed of extended family groups. But for Kropotkin, it was in tribes—tribes that he believed not to be bound by the artificial bonds of kinship—that mutual aid in humans originated and subsequently flourished. "Zoology and paleo-ethnology are thus agreed," Kropotkin concluded, "in considering that the band, not the family, was the earliest form of social life."[71] For Huxley, blood kinship was the one loophole in the gladiator contract that produced altruism; for Kropotkin, such kinship contributed little toward the ubiquity of mutual aid.

Not only did Kropotkin's time in Siberia embed mutual aid (in the absence of kinship) in his thoughts, it also made him an anarchist—a man who believed that the state stands in the way of what people do best: namely, live in small groups and help one another independent of blood ties. As Kropotkin saw it, "I lost in Siberia whatever faith in State discipline I had cherished before. I was prepared to become an anarchist."[72] In his obituary of Darwin, he mused that Darwin's ideas, when properly understood, were "an excellent argument that animal societies are best organized in the communist-anarchist manner."[73] It was the infidels, not Darwin himself, that were the problem—people who "reduced the notion of struggle for existence to its narrowest limits." These same infidels "came to conceive the animal world as a world of perpetual struggle among half-starved individuals thirsting for one another's blood. They made modern literature resound with the war-cry *woe to the vanquished*, as if it were the last word of biology. They raised the pitiless struggle for personal advantages to the height of a biological principle."[74] To Kropotkin, these were all completely foreign ideas: foreign and dangerous, both scientifically and politically. And Thomas Henry Huxley, of course, represented the personification of such dangerous ideas.

The altruism that Kropotkin saw in nature meshed smoothly with his political philosophy of anarchism, as is evidenced in his essay "The Scientific Bases of Anarchy." Anarchism in its most basic form promotes the argument that we need no centralized government to lead happy, just, and equitable lives. For Kropotkin, anarchists "arrive at the conclusion that the ultimate aim of society is the reduction of the functions of government to nil—that is, to a society without government, An-archy." Such a life arises when small groups work together freed from the yoke of central authority and the "the individual recovers his full liberty of initiative and action."[75]

Kropotkin believed that if animals, savages, and barbarians could undertake mutual aid in the absence of government, then surely civilized society needed no such regulating body and could live in peace, acting altruistically for the good of all and free from "the fetters of the State."[76] In promoting this view, he was, as an anarchist, simply following what he saw as the "the course traced by the modern philosophy of evolution. . . . [The anarchist] merely considers society as an aggregation of organisms trying to find out the best ways of combining the wants of the individual with those of co-operation for the welfare of the species." With this noble goal in mind, the anarchist "is thus a mere summing-up of what he considers to be the next phase of evolution."[77]

Anarchism was, for Petr Kropotkin, a natural solution to a natural problem. That is to say, it was not some human contrivance, but a fundamental outgrowth of processes that had been at work on creatures great and small for vast stretches of time. Altruism was an intrinsic part of nature, and it had nothing to do with blood ties. Kropotkin "gradually came to realize that anarchism represents more than a mode of action and a mere conception of a free society; that it is part of a philosophy, natural and social . . . that it must be treated by the same methods as natural sciences."[78] The natural science to which he turned was biology (natural history), and the philosophy he adopted was that which biology had taught him—the philosophy of mutual aid, of altruism. Mutual aid and anarchism were forever connected after Kropotkin's time in Siberia. Nature served as his guide for understanding and promoting the essentials of human kindness, for similar kindness and altruism prevailed in the animal world that he saw.[79] It was only the modern state, with governments that incorporate too much power, that Kropotkin believed was a danger to humans. When that happens, revolutions are sometimes called for, but only, Kropotkin argued, to bring us back to man's "natural state" and hence to mutual aid.

How could Petr Kropotkin and Thomas Henry Huxley—both of whom were held in high esteem in the scientific community of their day, both of whom were regarded as keen observers of nature, and each of whom had a strong following among the intelligentsia and the common man—see the natural world in such a dramatically different light? What made Huxley see nature as a bloodbath, with the one exception being family, but Kropotkin see a world of mutual aid divorced from any familial ties?

Part of the difference, no doubt, lay in their different life experiences. Kropotkin's was steeped in anarchism and social-ism, while in the Victorian England in which Huxley lived, com-petition was a mainstay; it was ingrained in everything British, from a world-class military that for centuries had dominated a good part of the planet, through the Adam Smith–like struggle in the workplace to the everyday life of schoolchildren, intensely jostling for a slot in the best schools to assure them of a powerful position in a hierarchical society. It was inevitable that compe-tition played a leading role in Victorian science and in the thoughts of one of its leading spokesmen.

But there was more to it than that. Kropotkin himself delin-eated one of the major reasons that he, and others in the Russian school of mutual aid, had such a different view of nature from Huxley and those in the English camp: "Russian Zoologists in-vestigated enormous continental regions in the temperate zone, where the struggle of the species against natural obstacles . . . is more obvious; while Wallace and Darwin primarily studied the coastal zones of the tropical lands, where *overcrowding* is more noticeable. In the continental regions that we visited there is a paucity of animal populations."[80] The same arguments that ap-plied to Darwin and Wallace applied to Huxley as well, if not more so. For the Russian evolutionary biology camp, led by Prince Petr, *environment* and ecology together were the key to understanding why mutual aid abounded.

The scientists—Darwin, for example—with whom Huxley associated spent their time studying life in the tropics, not in Siberia. In addition, Huxley himself had done a four-year stint as a medical doctor and naturalist on the HMS *Rattlesnake*, which sailed the tropics.[81] In contrast to the group living and group cohesion that was necessary for survival in Siberia, the life Huxley observed in the tropics was indeed more of a free fight, based on direct competition. There the struggle for exis-tence was literal, not metaphorical. Huxley saw the nature of life through the prism of the tropics, while Kropotkin saw a very different world in his arctic trek. In underpopulated Siberia, in-dividuals were living in a very harsh environment and aiding one another without any regard for blood kinship. Without such altruism, Kropotkin and his colleagues argued, they faced insur-mountable obstacles such as freezing temperatures and vast dis-tances between valuable, but scarce resources, and so, they were altruistic. For Huxley, the lush and crowded tropics epitomized intense competition, and there was but one refuge from this cannibalistic world: relations among family members.[82]

By any definition, Kropotkin and Huxley were among the intellectual giants of their era, and the public at large was fascinated with them. Kropotkin, for instance, spoke often and to anyone who would listen about his ideas on mutual aid and politics. Early on, his audiences were small. He thrived on conversations about mutual aid and politics in local pubs, meeting halls, or even in the fields, if that was where people were to be found. But as his reputation grew, so did his audience. In a short time he was recognized all over the world as a leader of the anarchist and socialist movements and the biologist behind the idea of mutual aid in animals and humans.[83] On his North American speaking tour in 1897, Kropotkin addressed crowds numbering in the thousands and did so again in a 1901 tour. Whether he was speaking on mutual aid or on anarchist-socialist politics (and sometimes it was hard to tell the difference) people came out in droves to hear him. Indeed, he was so popular that, in 1898, after negotiating some backroom deals, the *Atlantic Monthly* commissioned him to write a series of autobiographical essays.[84]

Huxley, too, reached a large audience and was honored for his work as a scientist and science writer all over the world, including Egypt, Sweden, Italy, Austria, Prussia, the Netherlands, and even Russia.[85] Wherever he went, he drew huge crowds. Indeed, at a "lay sermon" that he delivered one Sunday at London's St. Martin's church, two thousand people were turned away for lack of space.[86] Huxley could often been seen in characteristic suit, tie, and reading glasses. Before we became accustomed to our current vision of scientists in lab coats, people thought scientists should look exactly the way Huxley did—austere, brilliant, and somehow above it all. Indeed, Stephen Jay Gould, an "unabashed Huxley fan," argued that Huxley represented the first true modern scientist.[87] When he spoke, the audiences, be they intellectuals or those that attended his "Peoples' Lectures," paid close attention, almost as if they expected Professor Huxley to give them a quiz after the talk.

Both Huxley and Kropotkin felt that a fundamental understanding of altruism among living organisms was possible only through the study of evolution, with a particular focus on the role (or lack thereof) of family dynamics in the evolutionary process. Yet the argument can be made that Kropotkin and Huxley, scientists though they were, did not approach the questions surrounding evolution, kinship, and sociality in a particularly scientific manner. Neither of them ever formalized a theory on the connection between blood relatedness and altruism, let alone developed a mathematical model amenable to testing. Each just knew in his heart that he was right. They both translated their sense of purpose on these matters into statements that went far beyond what the science of their day could support.

What seems remarkable is the fact that so far not one of the characters in this drama did a single experiment examining altruism, cooperation, and kinship. Think about this sequence: Darwin's great theory floundered on kinship and altruism, two prominent nineteenth-century scientists battled on the subject, yet not a single experiment on the question had been generated—not one. To fault Huxley and Kropotkin for not constructing and implementing experiments, however, is to fall into the trap of gauging past actions by current standards. They did not undertake any experiments, not to mention controlled experiments, on animal behavior, because virtually nobody did so in the late nineteenth century. It was not until decades later that such experiments would be commonplace, culminating in 1973 when Niko Tinbergen, Konrad Lorenz, and Karl von Frisch won the Nobel Prize for their work in initiating the scientific study of animal behavior.[88] In the mid- and late 1800s, however, no one had laboratories for studying behavior or constructing controlled experiments on evolution and behavior. The structural framework for such things was simply not in place, and the study of animal behavior was essentially a natural history enterprise.[89]

Even so, we can ask what *might have* happened if any of these scientists had gone about his work by actually performing controlled experiments on the question of evolution and sociality. Take Kropotkin, for example. Like so many of his Russian contemporaries, the prince's views on evolution and behavior were drawn almost entirely from his observations of animals and humans in nature, observations colored by passionate political opinions of what society ought to look like. Fortunately, he had both the mental and physical strength to bear almost unthinkable hardship in order to amass his evidence. But what if, instead of devoting so much time and energy to the noble occupation of naturalist—of just observing nature in action—Prince Kropotkin had constructed some *experiments* designed to examine evolution, kinship, and cooperation?

It is possible to make an educated guess at the sort of experiments he might have done if he had cut short his Siberian journey and spent some time in the laboratory studying mutual aid, keeping in mind, again, that no such laboratories even existed in Kropotkin's day. Indeed, not only is it possible to venture a guess as to what might have happened; it appears that Kropotkin himself mulled over that question when he prophesized that "we must be prepared to learn some day, from students of microbial pond-life, facts of unconscious mutual support."[90] Our next character, Warder Clyde Allee, will fulfill the prince's prophecy and at the same time provide us with another glimpse into the personal side of the science of evolution, kinship, and altruism.

The Greatest Word from
Science since Darwin

Petr kropotkin published his book-length manifesto *Mutual Aid* in 1902. And while Kropotkin lectured on the subject for years after that, things were fairly quiet with respect to new work on altruism and its relation (or lack of relation) to kinship after the turn of the twentieth century. This changed in the early 1920s, when American ecologist Warder Clyde Allee[1] picked up where Kropotkin left off; only Allee was an experimentalist to the core. Like Kropotkin, he believed that altruism and cooperation lay at the heart of social behavior in nonhumans and humans, and that this cooperation was divorced completely from blood kinship. Indeed, he went so far as to suggest that the tendency to cooperate—with or without kin around—should be used to define life itself. But again, like Kropotkin, Allee's political, philosophical, and religious tenets regarding kinship and altruism made it difficult to discern when Allee the zoologist, and when Allee the devoted pacifist and Quaker, was talking. Although an experimentalist of great repute, he was anything but removed from the political and social ramifications of his work on altruism and kinship.

Like many other Quaker families in the Friends community that speckled the landscape of early-twentieth-century Indiana, the Allees traced their family history back to an emigration from one of the southern states—in their case, North Carolina. In a moral protest against slavery, many nineteenth-century Friends left the South for what they saw as more tolerant communities elsewhere. Parke County, the area where Allee was born and bred, was an active part of the Underground Railway,[2] and its beautiful fields and woodlands were a constant joy to the young Warder Clyde. Right until the time of his death, Allee described himself as just "an Indiana farm boy,"[3] and, as we shall see, ecology and environment played an important role in almost every aspect of his life.

Born on June 5, 1885, Allee spent his early childhood working on his family's farm and attending a one-room schoolhouse. An excellent student, he was admired by his fellow classmates for both his intelligence and his tenacity on matters that he felt were important. After high school he taught in another one-room schoolhouse near Rockville, Indiana, for one year and then taught sixth and seventh grades in Bloomington for a year.[4] At age nineteen, Allee enrolled at Earlham College, the local Quaker institution. This progression was typical for a boy with his background, as Quaker colleges of his day often recruited most of their students from the local community.[5] Like many young men who grew up in the rural world of Parke County, Allee had a fondness for nature and was particularly interested in ecology (a fairly new term in those days). Immediately after graduating Earlham, he enrolled as a Ph.D. student in the department of zoology at the University of Chicago.

When Allee began his graduate work, the University of Chicago was younger than he was, having been founded with the generous aid of John D. Rockefeller in 1892.[6] Nonetheless, his new academic home was already known as one of the premier American universities. In 1908, when Allee became a graduate student, the department of zoology was still under the auspices of the only chairman it had ever known, Dr. Charles Otis Whitman. Whitman's tenure lasted from 1892 until his death in 1910, and under his direction, research in zoology focused on embryology, with a particular focus on development of the individual organism from the single cell stage onward. Studies of behavior and ecology, on the other hand, were rare when Whitman served as chairman, which is rather odd, given that Whitman himself is often counted among the earliest experimental animal behaviorists.

From graduate student Allee's perspective, the department of zoology took a big step in the right direction when it named Frank Lillie to succeed Whitman as its head in 1910. By marrying into the wealthy Chicago family of plumbing magnate Charles Crane, Lillie had secured himself a position in the social circle of the Rockefellers and Carnegies.[7] This connection would be important in the development of the Marine Biological Laboratory at Woods Hole that Allee would come to love so dearly,[8] and it put Lillie in a strategic position from which to secure much-needed money for the department of zoology. Equally important, Lillie's zoology department took a turn toward ecology and behavior. With researchers such as Allee's supervisor, Victor Shelford, on board, physiological approaches to ecology and behavior soon became much more common at the university. Though the term ecology (or *oekologie*) was first coined by German zoologist Ernst Haeckel in 1869, the University of Chicago was one the few institutions that could boast of a program in "ecology"—defined as the systematic study of the interaction between organisms and their environment. This emphasis on how environment affected organisms would come to play a central role in the way that Allee thought about cooperation and kinship in both animals and humans.

In Allee's day, the University of Chicago was an institution that believed a biologist's job was to "discover nature's moral prescriptions and thereby serve as a savior."[9] So important was it to Chairman Lillie that the students at the university understand how the zoology department saw its role in society, that he wrote a rather incredible short article for the student newspaper, the *Daily Maroon*. In this piece, Lillie left no doubt as to the mission of the department: "The Department of Zoology is a research Department, all members of which are concerned in extending the bounds of knowledge. . . . The subjects of investigation in the Department of Biology furnish the basic scientific foundation for social control."[10] Going beyond even Huxley and Kropotkin, biologists at the University of Chicago were arguing that they could bring biological insight into human affairs—that the laws of nature, when understood properly, could help people in their everyday lives. Such was the world in which Allee was academically nurtured—a world where ecology was paramount and where it was expected that one would turn knowledge about ecology into something prescriptive, such as ideas on how to better shape society. And although this became Allee's view of the role of biology, one would not know it by looking at his graduate work.

Allee's dissertation research was a classic example of physiological ecology. His work centered on the factors determining the distribution of tiny crustaceans, called isopods, in different aquatic venues. These isopods would someday play a role in Allee's work on cooperation and kinship, but in his dissertation, he simply focused on isopod distribution in younger streams versus older ponds. Among other questions, he examined how an isopod's metabolic rate, itself a function of location and breeding season, influenced its responses to water currents. He tried to determine how metabolic rate could be influenced by the oxygen and carbon dioxide content of water and how metabolic differences emerged as a function of breeding season.[11] For Allee, the behavioral differences he found were not a result of genetic variation across isopod populations, but rather a function of *environmental differences*. At this early stage in his career, he was convinced that the environment and ecology of organisms were paramount in understanding life, and that this could be done with no reference whatsoever to heredity or the force of natural selection. This gravitating toward ecology and shying away from evolution and natural selection as forces driving behavior would carry over into his early work on cooperation and kinship.

Warder Clyde Allee received his Ph.D. in 1912, and on September 4 of the same year he married Majorie Hill. Hill was an English major who also spent time at both Earlham and the University of Chicago, and whose Quaker forebears, like those of Allee, had left the South to flee the evils of slavery. Majorie Hill Allee was hardly typical of the women of the early 1900s. In time she would become a well-known, award-winning author of children's books for girls;[12] over the course of years to come, the royalties she earned from her books became important for the Allee household. Majorie was Warder's friend and colleague in every sense, accompanying him each summer to Woods Hole Marine Biological Lab in Massachusetts, a summer refuge for many of the department of zoology's faculty. One of her most famous books, *Jane's Island*, although fiction, was based on their time at Wood's Hole.[13] She had turned her life with her husband into fiction once before in *Jungle Island*,[14] a book based on her adventures with Warder on Barro Colorado Island in Panama, including Warder's climb to the top of the forest's canopy, one of the first climbs ever made to the top of a forest of that size. From a purely practical perspective, Majorie's literary ability not only translated into royalties and a fictional account of her life with

Warder, but allowed her to review, edit, and comment on Warder's books and papers on cooperation and its independence from kinship.

In time, Allee and Majorie Hill would have three children: Warder Clyde Allee Jr. (1913), Barbara Hill Allee (1918), and Mary Newlin Allee (1925). Early family life for the Allees involved frequent shifts from university to university. But wherever they moved—from the University of Chicago to the University of Oklahoma to Lake Forest University back in Illinois—they enjoyed a peaceful home life, with Majorie as matriarch and a house filled with Quaker ideals, literature on social justice, and science books. To those, such as friend and biographer Karl Patterson Schmidt, who visited the Allees throughout their long marriage, "instead of contrasting," this juxtaposition "served to emphasize the unity of literature and science" to which the Allees truly subscribed.[15]

In 1921, Allee returned to the University of Chicago, where he accepted a faculty position in the zoology department. He would remain there for almost three decades. The position was presented as one that focused on "Ecology and Animal Behavior,"[16] and it was hardly unexpected that Allee should fill such a slot. Not only were his research interests a perfect fit (he was an ecologist interested in behavior), but "inbreeding" was all too common in the department, with many former students returning to take teaching positions. The zoology group thought quite highly of themselves and their students, and so who better to hire than one of their own?

Allee began his work on social behavior soon after he returned to Chicago. Over the next thirty years, he would write extensively on what he referred to as "cooperation." Just as Kropotkin's "mutual aid" was a catch-all phrase for what we would now call altruism, so did "cooperation" operate for Allee. His "cooperation" encompassed behaviors involving a cost to self in order to benefit others, but it also included other behaviors that involved helping others while in the course of helping oneself. To put his early work on cooperation and kinship into the proper perspective, it is necessary to understand one of the fundamental tenets of physiological ecology in Allee's day: namely, overcrowding always has deleterious effects on animals living in groups. By the early 1920s, there were dozens of papers, studying a wide variety of animals, that clearly demonstrated that overcrowding leads to stunted individuals, decreased rates of reproduction, "inferior" offspring, and a whole suite of other injurious effects at the individual and population level.[17] The implication of this work, with its Malthusian overtones, was that group life had a distinct downside.

Allee consistently began his early papers and books on cooperation not only by pointing out the overcrowding problem but by noting the commonly held belief that grouping behavior itself was inherently costly because "the same number of individuals are obviously more easily gobbled up by an enemy when aggregated than when scattered." In his opinion, "the older, grosser point of view dominated by the idea of a struggle for existence between animals" suggested that "aggregations are more obviously harmful than helpful." What this Huxley-like view of the world meant was that finding the benefits of group life became both "more difficult and more important."[18] Such benefits, however, were the lynchpin of Allee's work on cooperation and altruism, and he went to great lengths to document them. For what overcrowding and undercrowding did was shift the emphasis to the ecology of the organism and away from a focus on its evolution, and that change in momentum was of prime importance to Allee at the beginning of his career.

In his early work on cooperation, cognizant that overcrowding can have serious effects on individual and population function, Allee undertook experiments on the detrimental impact of undercrowding by studying "animal aggregations." He reasoned that in undercrowded conditions small groups might fail to obtain some of the benefits of mutual aid; indeed, there might actually be a definite *minimum* number of cooperators needed to obtain any benefits. To test this hypothesis, he first had to examine whether there were in fact benefits associated with group living, and then determine whether such benefits were affected by group size. If undercrowding was a problem, and grouping was hence beneficial, the possibilities for cooperation in such groups became real. And, in particular, if undercrowding was a indeed a problem that was solved by groups of cooperators aiding one another in the absence of kin bonds, then a more concrete theory for the existence and maintenance of cooperation might be within reach.

Allee spent long hours in his Chicago lab and at the Woods Hole facility running experiments, writing, and trying to establish the importance of cooperation for the ecology, physiology, and behavior of organisms. In a sweeping review paper in the *Quarterly Review of Biology*, and in his subsequent book *Animal Aggregations: A Study in General Sociology*,[19] he provided a litany of benefits for group living that would have made Kropotkin proud: benefits that increase with group size—up to a point. The rewards included increased life span, protection from predators, heat conservation, increased rates of reproduction, and the maintenance of water content in aquatic organisms. Modern animal behaviorists would couch these sorts of benefits in an evolutionary framework, but Allee saw them strictly in terms of the ecology of the organism. That is to say, modern theory ties ecology and genetics to natural selection, but for Allee—at least early in his career—the main question was whether organisms possessed those traits that matched them with their environment. He was not concerned with genetics, nor with evolutionary forces. And like Kropotkin before him, he believed that cooperation occurring in groups was divorced from blood kinship. The groups Allee was concerned with were, he believed, made up of unrelated individuals coming together to reap the benefits of cooperation. Blood kinship was beside the point.

While Allee was the first to give credit to the work done on cooperation in other labs, his own lab's work on group life and cooperation was both abundant and important. One example involves the little isopods for which he had developed a fondness in his graduate school days. Here, Allee tied together his interests in physiological ecology with his passion for understanding cooperation to test the hypothesis that individuals in groups of unrelated isopods will handle fluctuations in the availability of water better than isolated individuals. This should translate into the ability of individuals living in groups to better retain water in dry environments and better control water uptake in overly moist settings. If correct, this would suggest an immediate benefit to group life in the absence of kinship, as well as demonstrate one cost of undercrowding.

In one of his experiments Allee took ten groups of isopods and placed them on filter paper that would absorb any water that they failed to retain. Likewise, ten solitary isopods were placed on filter paper. After seven hours and forty-five minutes, all the isopods living in groups were thriving, whereas the lone individuals were dead from water loss. Allee was not able to pinpoint the precise mechanism by which group size modulated water uptake/output, but in one way or another, each individual living in a group lost about 16 percent of its weight, while the solitary individuals who died lost greater than 40 percent of their body mass. For Allee, delineating the mechanism controlling water loss was not as important as demonstrating the benefits of cooperation in groups, and the data clearly demonstrated that. Similar sorts of results were emerging from many different laboratories. Stick a group of ten green flatworms (*Convoluta*) in a suspension containing silver—which is poisonous to these organisms—and they can survive for forty-eight hours. Individual worms exposed to the silver will all be dead in a day's time.[20] Group living in these flatworms doubled expected longevity in the presence of poison.

Although Allee's experiments, such as those with isopods, were fundamentally simple in both structure and prediction, the results fueled his passion for studying cooperation and group life. "The formation of bunches helps make these isopods more independent of the water content of their surroundings and markedly decreases the rate of change of body moisture when this is out of equilibrium with their environment," he pronounced.[21] Dry as that summary may seem, Allee, like Kropotkin before him, believed he had demonstrated that the important element of the cooperation that emerges in group life is the endowment of individuals with abilities to better combat hardships in their environment. What really mattered was that the combat he witnessed was organism against *environment*, not organism versus *organism*, and also that kinship played no role in facilitating the cooperation that emerged when individuals joined together to fight the environment. Petr Kropotkin could hardly have said it better.

Allee garnered more support for his hypotheses from his own work on both starfish and sea urchins. He had observed that starfish living on the coast of New England were rarely found close together; in the wild, starfish lived a solitary existence and used the eelgrass in their environment to hide from predators. When he brought starfish into his laboratory, he found something quite different from what he had observed in nature. Starfish placed into small dishes containing only sea water, but no eel-grass, immediately formed large clusters. Individual starfish grouped so tightly that it was difficult to tell one individual from the next. If, however, Allee added artificial eelgrass to the starfish dishes, starfish aggregations dissolved, and individuals returned to their natural solitary state. He believed that this work demonstrated mutual aid among very simple creatures. When eelgrass was available, starfish were solitary because the grass hid them from predators. In the absence of such protection—that is, when the environment changed—cooperation was necessary, and starfish formed tight aggregations to ward off danger.[22]

Allee also cited his own work on the sea urchin as further evidence that cooperative behavior could be seen in primitive creatures. While working at the Marine Biological Laboratory at Woods Hole, he observed the behavior of *Arbacia*, a sea urchin common to the waters of Cape Cod. Fertilization in *Arbacia* is external; egg and sperm are released into the water, where they unite and begin the process of cell division. Allee measured the time to first cell division (from one to two cells), second cell division (from two to four cells) and third cell division (from four to eight cells), as a function of the numbers of fertilized eggs present. He found that the times to first, second, and third cell division were all accelerated when the number of fertilized eggs in his experiment was gradually increased from thirteen to more than four thousand. When grouped together, fertilized eggs also developed into free-swimming larva more quickly, providing him with further evidence that the benefits of group life and its consequent cooperation were not only ubiquitous, but ancient.

Allee believed that the benefits of group living in the absence of kinship were not restricted to times of extreme stress like those experienced by his experimental animals. Heat conservation, for example, was always an issue during winter, and animals in groups benefited from having lots of warm bodies around. What's more, many organisms produce substances that seem to "condition" their environment to make it more favorable to others of their own species in the vicinity, thereby opening the door for cooperation in the groups that might emerge. As a case in point, he cited A. Popovici-Baznosanu's work on the snail *Lymnaea*, in which individuals grown in water that had been occupied previously by other snails grew faster than individuals grown in pristine water.[23]

Allee was keen to note that the rate of individual reproduction is often higher in those living in groups.[24] His favorite example was T. Brailsford Robertson's work on reproduction in two species of infusorians (microscopic protozoans), *Enchelys* and *Colpidium*.[25] In both of these species, Robertson found that after a short delay, or "lag period," reproductive output (reproduction by cell division) in two infusorians was more than twice the rate of reproduction of a single individual. Indeed, the rate ranged from two and a half to ten times greater for pairs of individuals. For Allee, the conclusion was clear: animals living in groups of unrelated individuals profited through mutual aid. The benefits they reaped, he repeatedly pointed out, were subtle and omnipresent and potentially a much more important factor in animal life than the "struggle for existence,"[26] by which he meant a Huxley-like gladiator show.

As we have seen, Allee did not cast his work on cooperation in terms of evolution by natural selection, but instead he thought in terms of how environment shapes behavior. The fact that he did not initially think in evolutionary terms becomes less surprising when we realize that what is called the "modern evolutionary synthesis," wherein scientists from numerous disciplines came together to form a cohesive and powerful evolutionary framework for the biological sciences, was still almost two decades away when Allee joined the faculty at the University of Chicago. A lot of biologists—in particular, a lot of ecologists—did not place their thoughts in an evolutionary perspective. But there was a second, less scientific and more personal, reason that Allee initially steered clear of evolutionary thinking in his analysis of cooperative behavior. During World War I there had been a strong, palpable anti-German feeling among American scientists. Many German philosophers, as well as many German military leaders, interpreted "survival of the fittest" in terms of the military might of human societies, and this, in part, fueled a strong antievolution movement in the United States.[27] Allee's frequent exposure to this rhetoric, in conjunction with his own stance as a pacifist during the war, played a role in promoting his lifelong research program on cooperation.

During Allee's early days at Chicago he became obsessed with creating a cooperative world—a world that did not have evolutionary biology as one of its cornerstones. He rejected the social Darwinism most strongly associated with Germany and refused, at least at first, to take the chance that his research might be sullied by any association with this sort of thinking. At that time, for Allee the scientist and Allee the Quaker, ecology and environment alone would suffice in explaining and promoting cooperative behavior. He found it not only unnecessary to invoke evolution, but he intentionally steered away from it.

Allee's writings clearly show that his religious beliefs, including a strong devotion to pacifism, also led him to deemphasize the role of blood kinship in promoting cooperation. In essence, he was driven to study cooperation in order to find a biological justification against war; and, indeed, with time, he was convinced that he had succeeded. His commitment to this project was established early and appears to have solidified shortly after World War I, just before Allee came to the University of Chicago and began his work on aggregation and cooperation. Writing from the standpoint of a pacifist emotionally devastated by the war, he published articles on animal cooperation and its implications for understanding warfare in a wide variety of venues, from intellectual outlets like *Science* to political magazines such as the *New Republic* to local Quaker newsletters.

If he could show that cooperation was endemic in the animal world and that it was not tied to kinship, Allee believed he could debunk the myth that humankind was innately warlike and at the same time promote his firebrand version of pacifism. To understand just how tenaciously he pursued this goal, we need to realize that Allee was not just a pacifist, but also a regional leader in the pacifist movement. He was so well known for his convictions, that when *Time* magazine referred to A. J. Muste as "America's number one pacifist," Allee was asked by Alfred Watson, Midwest secretary of the Fellowship of Reconciliation, to host a dinner and fundraiser for Muste, an invitation that Allee humbly accepted.[28]

The expression of pacifist views was often distinctly unpopular in an era of world wars. For example, during World War I,

47

when Allee was a pacifist and conscientious objector, and held a seat on the Quaker War Service for Civilian Relief in Chicago, pacifists were openly ridiculed in public. Allee's local newspaper, the *Springfield News-Record*, learned of his position and mocked his support of pacifism, calling the conscientious objector law a "most convenient theory." Furthermore, the editors had their own ideas about what should be done with people like Allee: "Sometimes war is unavoidable and college professors are no more necessary to civilization than carpenters and cobblers. In fact, if we had to dispense with one or the other, we should prefer to give up the professors."[29]

While Allee's pacifism and his Quakerism were not the same thing, they were clearly linked, and Allee spoke as openly about the latter as the former. Although most modern-day biologists are not wont to discuss their thoughts on the relationship between religion and science, Allee felt that such an examination, particularly when it came to religious and scientific views on cooperation, kinship, and war, should be a primary driving force in a scientist's life. Thus, despite Allee's claim that he would be "very much surprised" if it turned out that his Quaker beliefs affected his "scientific activities and thinking," his own writings tell a different story.[30]

Take the case of his fascinating article "Reexamination of One Fundamental Doctrine in the Light of Modern Knowledge," which he wrote for a Quaker audience.[31] Here Allee faces head-on the relationship between war, science, and religion by examining whether war is ever justified. He believed that when confronted with arguments that war is in fact justified, "early Quaker Friends fell back on their inner feeling that war and its accompanying hatred are fundamentally wrong." But he was quick to note that "they had no proof of the correctness of their position on this fundamental point." And that is where he felt his experiments on cooperation among unrelated animals came into play. ". . . I have been interested in undertaking scientific experiments designed to throw light upon the fundamentals of this problem. . . . There is abundant evidence from modern science that the centuries old Quaker attitude of opposition to war is correct. . . . It is to our glory that we [Quakers] attempted at all times to substitute cooperation for struggle."[32]

Allee was always intellectually honest about his thoughts on cooperation, war, religion, politics, and science and would often preface such presentations with a caveat such as: "It is necessary to state that the present analysis is made from the objective viewpoint of science. Readers should be warned, however, that a long consideration of the problems involved has led me as an individual to take a somewhat extreme position as a pacifist."[33] Such statements garnered him respect from even his staunchest opponents.

Allee was clear that he wanted to develop a grand theory of cooperation among unrelated individuals—a theory that could be used to explain and promote cooperation in our species. But why study isopods and highlight work on reproduction in bacteria if what you really want is a comprehensive understanding of cooperation, especially in humans? Allee appears to have focused on these "simpler" creatures for both biological and social reasons. He realized that because of their small size and quick rate of reproduction, it would be easy to generate lots of data on such subjects; also, he believed that while most people might accept cooperation as integral to the life of "advanced" creatures, there was a reluctance among scientists and nonscientists alike to cast the net broadly. If he could show that simple creatures were constantly helping each other in the absence of blood kinship, this would surely strengthen his scientific claim that such cooperation operated virtually everywhere that life existed. From the sociopolitical perspective, Allee, like Kropotkin, believed that if cooperation without kinship could be proved ubiquitous in nonhumans, then humans, with their moral capacities, should be capable of achieving what seems to be an elusive state of being.

Simple cooperation among simple creatures, Allee believed, set the stage for more complicated types of cooperation in more complicated creatures. His work on simple creatures was merely the beginning of a sequence: "The first step toward social life in lower organisms is the appearance of tolerance for other animals in a limited space." Subsequent to this initiation of sociality, he held that "a first advance in social life is made when these groupings serve to promote the welfare of at least some of the individuals forming them." Next came mutual attraction—a force that brings cooperators together—and then group behaviors, in which individuals act in a coordinated, cooperative fashion. Allee was convinced that the scientific evidence for each and every step in this process was quickly being established and was there for all to see. And, also like Kropotkin, he was certain that all this cooperation was divorced from blood kinship.[34]

Allee continued his studies of cooperation from the early 1920s to his death in 1955, and he slowly began to incorporate more and more evolutionary thinking into his research. A key event in his progression toward an evolutionary approach to cooperation was Sewall Wright's joining the faculty of the zoology department at the University of Chicago in 1926.[35] Wright was an early advocate of what is known as "group selection" thinking[36] and would eventually develop the first explicit mathematical model of group selection in a paper entitled "Tempo and Mode in Evolution."[37] It was Wright's concepts that attracted Allee and sparked his interest in an evolutionary perspective on cooperation among unrelated individuals.

Although Wright's theories are extremely complicated mathematically, his ideas on group selection can be boiled down to the following model: even though some organisms may pay a price for cooperative or altruistic behavior, this sort of behavior can still evolve *if* groups with more (unrelated) cooperators outcompete groups with fewer (unrelated) cooperators.[38] For example, imagine a cluster of human groups. In some groups, individuals patrol the group's territory and guard against intruders and predators. Such guards are altruistic in the sense that they pay a price—they may be killed or injured during their duties—but everyone in the group benefits by their presence. If two groups start out at the same size but differ in the number of altruists (guards) within them, the group with many guards will, on average, have a higher survival rate per member than those with fewer guards. As such, group selection models predict that guarding behavior can evolve even though it has a potential cost associated with it.

Wright's models of group selection would later be modified and expanded by evolutionary biologists such as David Sloan Wilson and philosophers like Elliot Sober. In these "trait group" models, which emerged in the mid-1970s, natural selection operates at two levels: within-group and between-group. In the context of cooperation and altruism, within-group selection acts *against* altruists, since such individuals, by definition, take on some cost that others do not. Selfish types are always favored by within-group selection because they receive any benefits that accrue from the actions of altruists, but they pay none of the costs.

As opposed to within-group selection, between-group selection favors cooperation if groups with more cooperators outproduce other groups. For example, when individuals in a group give alarm calls, they pay a cost within their group, as they may be the most obvious target of a predator honing in on such a call. But their sacrifice may benefit the group overall, as other individuals, including other alarm callers, as well as those that do not call, are able to evade predators because of the sounded alarm. Thus, groups with many alarm callers may outproduce groups with fewer cooperators. For such group-level benefits to be manifest, groups must differ in the frequency of cooperators within them, and groups must be able to "export" the productivity associated with cooperation (for example, more total offspring or faster colonization of new areas). A heated debate still exists as to whether Wright's group selection model, or those that followed, can indeed explain the evolution of cooperation.[39] If such groups are full of blood relatives, however, almost everyone agrees that the debate over group selection disappears, as Wright's model becomes similar to modern kin selection models.[40]

In Wright's model, Allee saw a way to maintain his Quaker view of cooperation among unrelated individuals and at the same time incorporate an evolutionary perspective on his scientific work. He clearly believed that the strength of his own religious sect was in having many cooperators in its ranks, and this seemed to confirm in practice what Wright had demonstrated mathematically. Moreover, as an ecologist Allee was trained to think in terms of populations and groups, and here was an evolutionary theory that highlighted the ecology of the group. And it did so without any reference to the family group per se, which is to say that it was divorced from blood kinship.[41]

Another aspect of Wright's group selection theory that attracted Allee was the possibility that cooperative groups might fare better against their uncooperative neighbors by building up a larger population size. Groups with few cooperators, especially if they were isolated from other groups, might be smaller, and even go extinct—hence the benefit to moderate to large-size groups of cooperators.[42] This was an appealing part of Wright's theory, for it gave credence to one of Allee's first major insights— that underpopulation, under certain conditions, could carry with it a penalty. Indeed, in a letter to Thomas Henry Huxley's biologist grandson, Julian, Allee actually suggested that one could test this hypothesis on his very own Quakers: "... it has occurred to me that one might make an interesting study using human material. In some ways the small religious sects present phenomena not too far removed from the problem at hand and some of the Quakers, for example, have kept fairly adequate records. It may be possible to get some light on certain aspects of this problem then by combing the records of the small Quaker meetings."[43] He thought that from such records one could piece together the precise manner in which low population size might affect extinction of the group, and hence shed light on the power of Quaker cooperation.

Allee embraced group selection ideas not only because of the precision of Wright's mathematical analysis—much of which, by the way, was often too complex for most people, including Allee, to grasp in detail—but because of the work of another of his colleagues in the zoology department, Alfred Emerson. Emerson was the premier termite biologist of his, or for that matter, any time. So vast was his collection that when he donated it to the American Museum of Natural History in 1962, it contained samples from 93 percent of known termite species.[44] Emerson received his Ph.D. from Cornell University in 1925 under the tutelage of Harvard University's William Morton Wheeler, and he quickly adopted Wheeler's ideas on group selection.

In essence, both Wheeler and Emerson were drawn to group selection for the same reason that Darwin was drawn to selection at the level of the family: the existence of sterile insect castes. They, like Darwin before them, believed that this phenomenon only made sense if natural selection operated at a higher level than that of the individual. When it came to sterile insects, Darwin spoke of that level directly as the family (and to a lesser extent, the community). Wheeler and Emerson spoke of it more generally as the group level, but in essence they believed that Darwin was right when he pinpointed the most important group in the evolution of sterile castes as the family. Allee and Emerson were good friends and colleagues, and it was Emerson who convinced Allee that natural selection operates at the group level. Allee embraced this notion but simply rejected the idea that the critical group in group selection was the family. Rather, he felt that group selection was important in part because it drew attention away from individuals and families to the population itself.

Allee, again like Prince Kropotkin, was very much concerned that scientists and laymen alike understand that sociality, and hence cooperation, does not necessarily depend on family relations. Cooperation, he routinely insisted, can come into existence and remain a powerful force outside the constraints of the family unit. In his early publications, kinship is noticeable only in its absence; it is barely mentioned in his seminal 1927 *Quarterly Review of Biology* paper on cooperation and animal aggregations.[45] And when Allee did finally raise the issue of kinship and cooperation, he turned to the work of a seemingly unlikely spokesman on the subject—Herbert Spencer. Though often remembered for his extreme version of "nature red in tooth and claw," Spencer's early work focused more on altruistic and cooperative behavior. In his *Animal Aggregations* book, Allee notes that "Herbert Spencer's suggestion that colony life arose from the consociation of adult individuals for nonsexual, cooperative purposes was an early recognition of that type of social unit."[46] He is quick to note, however, that at the time they were advanced, Spencer's ideas were "not well grounded on proved fact," but that since then, much evidence had been amassed in support of the notion that many, if not most, cooperative groups did not owe their origin or current makeup to blood kinship. As Allee saw it, Spencer's ideas "may be correct in many instances."[47]

In his 1927 book, *Animal Aggregations*, Allee compiled pages of examples to support the hypothesis that sociality did not generally rest on, or revolve around, the family unit. For Allee, the group was of paramount importance, yet the *family* group per se played only a limited role in the cooperation he saw in most species. Later he hedged his bets a little and noted that "the extension of family relations is very obviously one potent method by which social life is developed to a higher level," but even on such rare occasions, he would inevitably and almost immediately add a critical caveat such as "there are other methods that also deserve consideration."[48]

Allee believed that for most life forms, cooperation was not structured by the family, the one partial exception being "highly integrated societies," by which he primarily meant birds, mammals, and social insects. But as always, he would then demur that even in these taxa, many instances of cooperation are divorced from family life. More important, he claimed, was the fact that "at least a part of the forces which operate to bind members of families together are similar to those which operate in the case of other kinds of social groupings."[49] In other words, even when families played a role in the evolution of cooperation, it was subordinate to the general forces that work on all groups, not just families alone.

Warder Clyde Allee's ideas on sociality and kinship were anything but myopic and did not revolve only around the work coming from his own lab. He understood these questions to be both important and general. Broadly speaking, he was convinced that his message that cooperation and kinship were not conjoined was one that must be vigorously promoted to counter the flow of opinion in the opposite direction. Though the issue did not reach the level of importance in his day that it would in years to come, some biologists were already talking about the topic, and usually arriving at conclusions that were counter to Allee's. The most potentially unpalatable of these ideas to Allee was the application of his opponents' ideas to *Homo sapiens*, in that sociologists and anthropologists were tracing the roots of human society to the family.[50]

Allee did not mince words in his response to the idea of the omnipotence of family as the explanation for such social behaviors as cooperation and altruism in nonhumans and humans. He strongly believed that the group per se, *not* the family group, was the fundamental unit in which cooperation and altruism evolved in nonhumans. And when it came to human life, kinship, and family, the case was equally compelling, and laid out again by none other than Herbert Spencer, who argued, ". . . In terms of human society . . . the gang, rather than the family, [w]as the preliminary step in the evolution of the social habit."[51]

His position on kinship and cooperation did not come without a price, for his ideas put him in direct conflict with the leading anthropological theory, which focused on the family unit.[52] One serious penalty that Allee paid for his views on kinship came in the form of a key currency in the academic world—grant money. Allee was chronically underfunded, but not for want of trying. The problem he faced was that one of the major funding sources that the University of Chicago faculty relied on was the Rockefeller Foundation and its subsidiaries, a group heavily populated by social scientists who adopted the anthropological stance that kinship was the key to understanding behavior. A good example of this state of affairs can be seen in his interactions with the National Committee for Research in Problems of Sex.[53] This organization, whose work extended into research on kinship, consistently funded Allee's colleagues such as Raymond Pearl and Frank Lillee. Allee could have quite easily tweaked his research program to fit within the bounds of what the committee deemed worthy of funding, but he refused to do so, as it would have meant being unfaithful to his most basic tenet—that cooperation could be found virtually everywhere and was not tied to the family—and damn the scientific establishment if they could not see what was obvious to him.

Part of Allee's obsession with downplaying the role of the family in the development of cooperation appears to be rooted in his Quaker outlook on societies in general, and human society in particular. He believed that to make the world a better place, we need to break down, not build up, barriers between people—indeed, if necessary, even by supporting the notion of a world government.[54] He was adamant that "the practical method of securing a broader base for cooperation, whether it be inter-racial, intersectional or international, lies in exposing children of different races, . . . or nations to each other during early impressionable years when thought habits are being formed. Then if they become thoroughly accustomed to each other, new cooperative units will develop. . . . In this respect, human nature in essentials seems not that far removed from ant nature or from animal nature in general."[55] In Allee's utopian society, artificial boundaries, including gender, nationality, race, and *family*, needed to be removed.

Allee believed that people should treat *every* decent person as if he or she *were* family. But that hardly seemed to be the way things work, and hence the family unit as a structure, which most often divided populations rather than united them, was not the place to look for universal goodness. In fact, Allee often wrote about artificial boundaries that strangled what he saw as mankind's capacity for cooperation. Rather than delineating "us" from "them" at a whole series of levels, Allee's quest was an all-inclusiveness, one that cast off the shackles that heretofore limited our amity to one another. As he saw it, history showed that people tended to see whatever group they belonged to as being the one that evolution favors, and hence they consistently opposed any notion of expanding the rights attained by their group to a larger membership. He seemed to accept this as a fact of life, and noted (half in jest and half seriously): "So might the conservative primitive-living molecules, the protozoans, the flatworms, isopods or ants have argued, had they the wit, at each stage of their cooperative evolution."[56] Fact of life on this planet or not, he was determined to make cooperation, without kinship, the ideal to which mankind should ascribe.

Allee's Quaker views led him to look for models of goodness that did not rest on the family. "The birth of society through the family," historian of science Greg Mitman notes, "posited a sexual division of labor, a patriarchal hierarchy in which the male, bearer of culture, ensured stability and order by keeping nature in her place. The female, constrained by biology and thus bound to nature, was destined for a life of reproduction and child rearing."[57] This certainly was not the way it was in the Allee household. But more important, this hierarchical vision was in direct opposition to Allee's political leanings. He saw biology as a means of promoting a society devoid of hierarchy, one in which cooperation and pacifism were the law of the land. Cooperation via family relations—the birth of society through the family—would not get him to where he wanted to go.

Allee believed that in its entirety, the results of his and other scientists' work on cooperative behavior suggested that nonkin cooperation was a deep-seated rule of nature. He argued that cooperation, not aggression, was selected for in animals, and, by extension, in humans; indeed he went so far as to say that "those who assert that the whole trend of science is to lend support to the present war system in settling international disagreements are relying on a mistaken, outmoded phase of biological thought."[58] The study of cooperation not only lay at the heart of his scientific agenda; it was a mission, and he had every intention of proving that Thomas Henry Huxley was dead wrong when he argued that man should not turn to nature for the answers to questions about morality. Allee's research program, in conjunction with his extreme position on pacifism, led him to make some incredible claims regarding the power of cooperation, one of which focused on the definition of life itself.

In 1953, Allee was asked by editor James Newman to contribute a section to a volume on modern scientific thought that Newman was putting together for the publishing house of Simon and Schuster.[59] Allee's chapter would cover "Ecology and General Biology," while interestingly enough, his friend, Thomas Henry Huxley's grandson Julian, would have the honor of writing the section on "Evolution." Allee's essay in this subsequently well-read volume was entitled "Concerning Biology and Biologists" and contains a clear caveat for the reader: "In order that my possible personal bias, if any exists, may be apparent, I should say that I am a citizen of the United States of America, with generations of American ancestors. Further, I am both a mature biologist, and a working, though highly unorthodox member of a religious organization."[60] As always, despite his somewhat extreme positions on many issues, he maintained an air of intellectual honesty.

Allee devoted many pages in his chapter to a discussion of the origins of life and asked: "How did life originate? . . . We actually know little of how life began. . . . This was not always so. I knew in my boy-hood on a back-country Indiana farm that life had been created by a kindly, very wise, so capable, 'Mr. Jehovah' and that was that." Allee recognized that in order to get a handle on the origin of life he need a working definition of it, and in this context he made one of his most far-reaching claims about cooperation. In a list of the attributes that constitute life, Allee included the fairly uncontentious claims that "all can reproduce their kind" and "all are continuously adjusting to their environment." The last item on the list, however, is "all show at least the forerunners of cooperation."[61] So ingrained in Allee was the idea that cooperation was omnipresent, that he felt comfortable including it on a checklist of the properties something must show in order to be considered alive. Indeed, in other writings he goes even further: so deep does this principle run that "the mutual dependence of the living must have grown out of similar but simpler interdependence in antecedent non-living matter."[62] Julian Huxley's grandfather was probably spinning in his grave after that line.

With such a strong opinion on cooperation, it is not all that surprising that in publishing his papers and books, giving seminars, and trying to convince the powers-that-be that his work should be funded, Allee often took on the role of a salesman. For example, at the start of his book *Cooperation among Animals*, Allee presented his strongest pitch for cooperation: "Widely dispersed knowledge concerning the important role of basic cooperative processes among living beings may lead to the acceptance of cooperation as a guiding principle both in social theory and as a basis for human behavior. Such a development when it occurs will alter the course of human history."[63] For Allee, nonkin-based cooperation was mankind's best hope for survival.

As any good salesman knows, to get your product to market requires both capital and air time. Allee knew that his battle for money was going to be a hard, uphill one. In his notes from a National Research Council (NRC) meeting where he (along with other zoologists in his department) was trying to solicit funds for his work on cooperation in the absence of kinship, he wrote: "Make it clear to people with money that the work of this committee is a good investment. Former statement . . . not strong enough."[64] In addition to a struggle for capital, Allee could see that spreading his message about nonkin cooperation was going to be difficult as well. He found his responsibility to promote the importance of biological work on cooperation was all the more difficult to fulfill "since there is a desire for sensational developments and sensational points of view to the extent that the more calm, orderly presentations do not reach the ear of the general public to the extent that many of us would like."[65]

In the eyes of some people, particularly those who controlled the funds that Allee sought, his salesmanlike approach to the study of ecology and cooperation, and the profound impact such work had on human society, was overkill. For example, Allee was a key player in an attempt to create a Committee on Population Studies at the NRC. If such a committee was created, it would fund just the sort of work that Allee and his colleagues were undertaking. On February 19, 1942, as the United States was entering World War II, Allee and his colleagues submitted a long report to Robert Griggs, one of the chairmen of the NRC, in which they outlined the role for the proposed Committee on Population Studies. Two of the primary purposes of such a committee were "a developing interest in comparative studies among students of human sociology" and the "discovery of nonconscious cooperation among animals."[66] Given this, Allee, for his part, made sure that the committee's report contained strong statements on the essential role of nonhuman cooperation studies in guiding human society. His grant was rejected and then put on hold for the remainder of the war, after which time it surfaced once again in the form of various new proposals to create an NRC Committee on Population Studies.

The arguments in Allee's grant proposal, which were often copied almost verbatim from his books and journal articles on co-operation, were more than the NRC's Griggs could tolerate, and he responded on two levels. In a series of letters in 1947, Griggs protested that Allee was extrapolating far beyond reason, and that it was getting a bit irritating: "It is fine for you to say that the study of [animal] population problems is the key to establishing the peace of the world. . . . If you could prove that, there ought to be loads of money to help you do the work. But as it stands now, there seem to be too many links in the chain of reasoning con-necting research on animal population and peace of the world." To add insult to injury, Griggs then informed Allee that not only did the data not support his contention but that he was not a par-ticularly good salesman to begin with: "Recently I submitted your report . . . to someone who knows nothing about ecology. He reported that it was very interesting reading . . . but that it did not constitute much of a sales talk and wouldn't induce anyone to give money for the cause."[67] Most of the time Allee's writings and speeches were received in a very positive light, but the Griggs incident demonstrates that his ideas on cooperation be-tween unrelated individuals, and the way he pitched them, were hardly universally accepted.

Mandatory retirement laws at the University of Chicago forced sixty-five-year-old Allee to retire in 1950, but he was quickly offered the position of chairman of the department of bi-ology at the University of Florida, which he accepted and held for the remainder of his life. His last five years were relatively uneventful, with one notable exception. In 1951, based primarily on his work dealing with cooperation in the absence of kinship, Allee received one of the most coveted awards any scientist can hope to achieve: election to the National Academy of Sciences.

On March 18, 1955, two days after admission to the hospital with a kidney infection, Warder Clyde Allee, passed away. At his funeral service, John Stewart Allen, president of the University of Florida, quoted an unnamed sociologist he knew who noted that "Dr. Allee has brought us the greatest word from science since Darwin. His scientific research which shows the evidence for cooperation gives us new hope."[68] Though the first sentence would have embarrassed the usually soft-spoken Allee, the second one is exactly the way he wanted to be remembered. Since his graduate student days at Chicago, where they emphasized that "the subjects of investigation in the Department of Biology furnish the basic scientific foundation for social control," Allee saw his work on cooperation as a beacon of light. Biology, could, he was certain, draw people in and teach them to be cooperative—if they would only listen. This, of course, was also the mission of Warder Clyde Allee, devout Quaker and pacifist.

As we have seen, during the latter part of his career, Allee relied on his department colleague, Sewall Wright, to provide him with an evolutionary underpinning for his own work on cooperation among unrelated individuals. Allee chose his friends well, as Wright, along with Ronald Fisher and J.B.S. Haldane, is considered one of the founding fathers of the modern approach to genetics and the study of evolution. But Wright's views on nonkin altruism were his own and not shared by either Fisher or Haldane. Indeed, Allee, who spent his career championing nonkin cooperation, would have been shocked and dismayed if he had ever heard the offhand remark about kinship, evolution, and altruism once made by Haldane—the next player on our stage.

J.B.S.: The Last Man Who Might Know All There Was to Be Known

The one great difference between man and all
other animals is that for them evolution must always
be a blind force, of which they are quite
unconscious; whereas man has, in some measure
at least, the possibility of consciously controlling
evolution according to his wishes.
—*J.B.S. Haldane and Julian Huxley*[1]

J.B.S. HALDANE was a man fond of one-liners. Scientist, science writer, and science fiction writer wrapped into one, Haldane once summarized his views on life by noting, "My own suspicion is that the universe is not only queerer than we suppose, but queerer than we can suppose."[2] Legend has it that when asked about our own little planet, and more specifically about what evolutionary biology might tell us about God, Haldane cracked that God must have "had an inordinate fondness for beetles," which seems a rather strange thing to say, until you realize that there are over 350,000 known species of beetles on the planet, and probably many more still unknown.[3]

But it was another of Haldane's off-the-cuff remarks that would mark the start of the modern mathematical theory of kinship and altruism—something that was sorely needed, as no such theory existed in Haldane's day. Haldane was keen on telling people that he would jump into a river and risk his life to save two brothers, but not one; and that he would do the same to save eight cousins, but not seven. Using himself as an example, the point Haldane was trying to make was that the more closely related two individuals are, the greater the probability that one will sacrifice for the other. That is, if we know something about the genetics of blood kinship, we can make predictions about the amount of altruism that will occur. In retrospect, it seems obvious that if kinship matters in selecting for altruism, then the degree of kinship should matter as well. But that was far from obvious in Haldane's era—indeed, he appears one of the first to have even thought about the question this way.

Haldane was considered by many to be a true genius—a man equally at home writing complex mathematical models or quoting from Dante, the Old Testament, or Sanskrit Hindu epics.[4] Science fiction writer Arthur C. Clarke described Haldane as "the finest intellect it has ever been my privilege to know,"[5] and one of Haldane's students summed up the reverence most who knew him felt by noting that Haldane seemed to him "to be the last man who might know all there was to be known."[6] A brilliant mathematical geneticist, as well as a popularizer of science, Haldane seems to have stopped short on his path to formalizing his ideas on kinship and altruism, limiting himself to a few passing, albeit important, remarks. Why he did so is something of a mystery, for Haldane was, in all other respects, one of the most prolific scientists of his time, publishing more than three hundred science articles, dozens of books, and hundreds of popular pieces, including articles in the *Manchester Guardian*, the *Daily Mail*, *Atlantic Monthly*, *Harper's*, and the *New Republic*.[7] Whatever the reason he abandonned this subject—and we shall return to some interesting speculation on this matter later—his thoughts on evolution and altruism set the stage for much of what is to come in our journey into kinship and the evolution of altruism and cooperation.

John Burdon Sanderson Haldane was born on November 5, 1892, in Oxford, England. If Thomas Henry Huxley, one of Haldane's intellectual heroes, began a family intellectual dynasty, Haldane inherited one; and as we shall soon see, these dynasties intermingled in unexpected ways. Haldane's father, John Scott or J. S. Haldane, was an early member of the Eureka Club, a collection of brilliant Edinburgh scientists and naturalists that would eventually contribute seven members to the Royal Society.

Haldane's father was kind and courteous, but also a tough man who would not put up with nonsense or arbitrary action on the part of authority. Indeed, when describing his father, J.B.S. Haldane noted that "he was born with a historically labelled Y chromosome. That is to say, his ancestors in the putative male line since about A.D. 1250 are known. There are, I believe, about fifteen similarly labelled sets of Y chromosomes in Britain. Their possession is generally a handicap, but may help protect the possessors against the voice of Establishment."[8] John Scott Haldane's early work on air in houses, sewers, and schools quickly garnered him a reputation as one of England's leading physiologists, and a man who was often called on by the government to assist in matters of national defense. J. S. Haldane's work in physiology, for instance, had led to the development of the Admiralty Diving Tables, immensely important in terms of protection against decompression sickness for divers.

As a young boy, J.B.S. Haldane was both stubborn and precocious. By age three he could read, and legend has it that before he turned four, upon discovering blood on his forehead, he inquired, "Is it oxyhemoglobin or carboxyhemoglobin?" At age eight, he fell off his bicycle and received a serious fracture of the skull. While at the hospital, his father told the authorities that his son had "a fracture of the base." J.B.S., unconscious until this point, opened his eyes and asked, "Base of what?"[9] The doctors told John's parents that the damage the boy suffered was life-threatening and that should he survive, he would likely be mentally retarded. But Haldane recovered perfectly, quickly regaining his strength, both intellectual and physical. Among his friends the tale emerged that somehow the brain injury he suffered turned a brilliant young man into a genius. Rumors and speculations about the beneficent outcome of a tragic accident aside, part of J.B.S.'s precocity was no doubt due to the fact that science, and more to the point, scientific experiments were a regular part of life in the Haldane house.

When he was eight years old, J.B.S. was already assisting his father with experiments on measuring gases, calling out numbers and calculating the amount of different gases present in the mines. This was very difficult work, but J. S. Haldane expected much of his son and would often use quite unconventional means to teach him a lesson. J.B.S. recalled an incident from his youth when he and his father were working together in a mine: "To demonstrate the effects of breathing firedamp, my father told me to stand up and recite Mark Antony's speech from Shakespeare's 'Julius Caesar,' beginning 'Friends, Romans and Countrymen.' I soon began to pant, and somewhere about 'the noble Brutus' my legs gave way and I collapsed on the floor, where, of course, the air was alright. In this way I learnt that firedamp is lighter than air."[10] That same year J.B.S., also through the courtesy of father, had his first exposure to what would become one of the loves of his life when John Scott Haldane took his young son to hear A. D. Darbishire's Oxford University seminar on the rediscovery of Mendel's laws of genetics. In time, this exposure would set the stage for Haldane's laying out the first scenario that tied altruism and kinship directly to genetics. But that was still years off.

With John Scott as his father, it was only natural for a boy of J.B.S.'s standing to attend the elite and prestigious Eton School, in preparation for attendance at Oxford University. Most of the students at Eton did not like Haldane, in large part because even in a school of talented youth, he stood above the crowd. Although he was a big young man, he was picked on constantly, and this experience ingrained in him a steadfast sympathy for the underdog in any contest. The notable exception to the Eton students' hostility was Thomas Henry Huxley's grandson, and Warder Clyde Allee's friend and colleague, Julian Huxley, whom Haldane remembered for presenting him with an apple, a ritualistic mark of friendship at Eton.[11] The apple (of that meeting) marked the beginning of a lifelong friendship that, on the intellectual front, produced a collaboration on a major basic biology textbook.[12] While at Eton, despite a proclivity for science, Haldane displayed what would become a characteristic pattern: he switched from one specialization to another, backed by his father, who rejected the accusation of the headmaster, Canon Lyttelton, that his son was becoming a "mere smatterer."[13] The result was a course of study that was a mixture of classics, chemistry, physics, history, and biology.

Following his days at Eton, Haldane attended Oxford University's New College, where he learned more of Darwin's and Mendel's work at the hands of a fascinating character named Edwin S. Goodrich.[14] Oxford provided a much needed respite from the social problems that Haldane experienced at Eton, and there he quickly made friends with Julian Huxley's brothers, Aldous and Trev, as well as the Gielguds. Though quite interested in math, Haldane claimed that "no one could study mathematics intensively for more than five hours a day and remain sane,"[15] and he again proved himself to be an academic omnivore and studied everything, finally settling on a degree not in mathematics or science, but classical literature, or what was known as "Greats."

At Oxford, Haldane joined the Officer's Training program, and in July 1914, he volunteered for service in World War I, requesting and eventually receiving a commission in the famed Scottish unit, the Black Watch. He quickly became an officer in the First Battalion in France, and despite the horrific bloodshed of World War I, J.B.S. enjoyed the experience. Indeed, he seems to have taken pleasure in both attacking and even being attacked by the Germans, earning himself the moniker "Bombo" along the way.[16] In a latter-day essay, "Illnesses That Make Us Healthier," Haldane goes as far as to say that "he enjoyed the opportunity of killing people" and regarded this "as a respectable relic of primitive man."[17] One remarkable aspect of Haldane's World War I experience was, as he would later boast, being "the only officer to complete a scientific paper from a forward position of the Black Watch."[18] This paper, published in the *Journal of Genetics*, was the outcome of the experiments that Haldane, his sister Naomi, and A. D. Sprunt had begun during Haldane's days at Eton.[19]

While at Eton, Haldane had become fascinated with genetic linkage, and he and his colleagues undertook a series of experiments examining this phenomenon in mice. Genetic linkage is said to occur when *a set* of genes appear to be passed down across generations as a unit.[20] Usually this occurs when the genes in question reside near one another on a chromosome and are inherited as a unit rather than piecemeal. What Haldane and his colleagues found was that two genes in mice appeared to be inherited as a single unit. More specifically, in mice one gene (labeled factor C) controls albinism or its absence, while a second gene (factor E) codes for pink eyes or normal dark eyes. By breeding various genetic combinations, Haldane's group was able to show that the gene for albinism and pink eyes were linked, with pink-eyed albinos occurring much more frequently than would be expected if these two traits were inherited independently of one another.

Haldane's collaborator on the mouse work, A. D. Sprunt, had been killed in action, and J.B.S. wanted to assure himself that should he too die in the course of the war, this work would be published. This paper was only the second article demonstrating linkage in animals ever published, and Haldane delighted in the fact that he had no formal degree in biology, let alone genetics. In the affiliation below his name at the start of the paper, "Lieut, 3rd Black Watch" ceremoniously preceded "New College, Oxford."

After World War I, Haldane returned to New College, Oxford, this time as a demonstrator in physiology, which on the surface of it appears absurd, because he had no degree whatsoever in that area. Yet, here, like at many pivotal points in his life, he used his father—who never took a course in engineering but went on to be president of the Institution of Mining Engineers—as a model. Despite the fact that Haldane was "a terrifyingly bad experimenter,"[21] he and his colleague Peter Davis began testing some of his father's ideas on the role of carbon dioxide in the bloodstream, as well as on the role of blood alkalinity on breathing. Once again, with his father's approach to science as a model, Haldane insisted on using himself as a test subject, adopting an attitude that "you cannot be a good human physiologist unless you regard your own body, and that of your colleagues, with the same sort of respect with which you regard the starry sky and yet as something to be used and, if need be, used up." In his own experiments, Haldane took extensive notes and referred to himself in the third person: "J.H. panting . . . ," "J.H. finding difficulty in breathing. . . ." In another experiment on the acidity of blood, he drank eighty-five grams of calcium chloride over thirteen days and noted "intense diarrhoea, followed by constipation due to the formation of a large hard faecal mass."[22]

In 1925, Haldane left Oxford for a much more stable position at Cambridge University's Trinity College, to fill a new readership in biochemistry. This appointment would allow him to pursue his interests in understanding the biochemistry of the gene. In the 1920s, the demarcations between genetics, biochemistry, and physiology were not nearly as distinct as they are today, and as Haldane's colleague, Norman Pirie, noted, "The only surprising thing would have been if he [Haldane] had moved into geology or astronomy, and either would have been possible."[23] In fact, almost to prove the point, in 1930 Haldane became Royal Institution Fullerian Professor of Physiology at Cambridge, succeeding his friend Julian Huxley. By this point, Haldane had worked in genetics, physiology, and biochemistry, but had expressed no real interest yet in anything to do with the evolution of altruism and its relation to blood kinship.

That would change, but not for another couple of years. Shortly after arriving at Cambridge, he met the woman who would become his first wife, Charolette, a writer for the *Daily Express* who was planning to write a novel—a science fiction piece called *Man's World*. She wanted to talk to Haldane about the science she was discussing in her book and convinced her editor that this would make for a good story. Haldane talked with her, but asked to remain anonymous. "To my astonishment the resulting paragraph in the Daily Express not only kept to the facts, but, as I had stipulated, did not mention my name. For this, and other reasons, I fell in love with the reporter and my love was reciprocated."[24] Born Charolette Franken, the woman who reciprocated Haldane's love was at the time Mrs. Charolette Burghes, and she and Haldane would have to wait to marry until after she obtained a divorce.

Divorce was no slight matter in England in the 1920s; indeed, adultery often had to be proven in court, in nasty detail, for a divorce degree to be granted. Haldane and Burghes arranged for just such detail by audaciously hiring a detective to follow them into a hotel. As Haldane recalled, "The next morning he appeared in our bedroom with the morning papers. Save for one moment when I had feared that we might lose sight of the detective, everything passed without a hitch."[25] Well, almost everything. After Mr. Burghes filed for divorce, Haldane and Charolette took the extraordinary step of saying that they had not committed adultery, which apparently was technically true, as there was no direct proof of sexual infidelity. The judge, Lord Merrivale, however, would have none of that, and awarded Mr. Burghes a divorce and fined Haldane one thousand pounds, damages to be paid to Mr. Burghes.

The fireworks regarding his marriage to Charolette, however, had just begun, for now Haldane had to deal with his colleagues at Cambridge—colleagues such as Professor William Bateson, who ran the Horticultural Institute, who on learning of Haldane's actions noted, "I am not a prude, but I don't approve of a man running about the streets like a dog."[26] Many others at Cambridge concurred, believing that Haldane had displayed "gross or habitual immorality"—grounds for dismissal. The case against him was brought before a Cambridge tribunal called the Sex Viri, which translates as "six men," but Haldane delighted in the fact that the Latin pronunciation of *sex viri* is "sex weary." Haldane fought the charges brought against him, and after a number of skirmishes secured his position at Cambridge.[27]

It was during his first ten years at Cambridge that Haldane became one of the most respected scientists in England, primarily as a result of a series of ten brilliant papers on the theory of evolutionary change published in the journals *Proceedings of the Cambridge Philosophical Society, Transactions of the Cambridge Philosophical Society,* and *Genetics.*[28] These papers, each of whose titles began "A Mathematical Theory of Natural and Artificial Selection . . . ," developed a theoretical framework for predicting evolutionary change under a wide number of conditions—different population structures, different systems of genetics, and so on. To a large extent, these articles led to most of the honors that would be bestowed on Haldane over the years, accolades such as the Darwin Medal, the Legion of Honor Medal (France), the Huxley Memorial Award, the United States National Academy of Sciences' Kimber Medal, as well foreign membership in the Royal Danish Academy, the Humboldt Academy of Sciences, the U.S. National Academy of Sciences, and the U.S.S.R. Academy of Sciences.

To understand why Haldane's models were so important in the 1930s, we need to realize that at the time, a debate on the nature of evolutionary change was raging in biology. Either natural selection was acting in a slow, methodological way on small differences as "selectionists" claimed, or via *saltationism,* that is evolutionary change via large-scale, sudden changes from the norm as so-called Mendelians claimed.

Mendelians, who were primarily lab-based geneticists, were trained more as physical than natural scientists. They held that most variation seen in organisms in nature was caused by the environment and was largely not genetic in origin. What's more, for Mendelians, most of this natural variation was on a small scale and not all that important for understanding real evolutionary change; it was akin to noise in a system. Instead, they conceptualized evolution as involving dramatic, large-scale—saltational—changes.

In the other camp were the selectionists or naturalists, who primarily worked in the field and were enamored with the huge amount of variation they saw all around them in nature. For selectionists, it was the small-scale differences—what the Mendelians thought of as noise—that was the stuff upon which natural selection acted. This group, the direct intellectual descendants of Darwin, saw selection as a gradual process. One offshoot of this debate was that it got scientists talking about genes for different traits: genes for morphological traits, genes for anatomical traits, and particularly salient for our discussion, genes for behavioral traits like altruism. And if, such genes existed, then one question becomes: how could population structure—kin-based or not—affect selection of such genes?

The problem early on was that there was no empirical or theoretical work that really could distinguish between the claims of the Mendelians and the selectionists. As the empirical work began to flow, it was becoming clear that it supported the selectionist view of evolutionary change. But there remained no mathematical theory that explained how a slow gradual process like that argued for by selectionists could operate. Haldane's models changed all that, as they showed precisely how natural selection operating in a slow methodical fashion could produce significant evolutionary change—even, it would turn out, when the trait in question was altruistic behavior.

Haldane's series of mathematical models on evolutionary change marks a shift from actual physiological experiments to "thought experiments" and mathematical models. No longer a bench scientist, Haldane moved into the world of numbers and the universe of theory. Along with Ronald Fisher and Sewall Wright's work, Haldane's models also mark a turning point in evolutionary biology—a shift from the field and the bench to paper and pencil.

For J.B.S. Haldane, 1932 was a particularly good year at Cambridge. He was elected a fellow of the Royal Society, and his now classic book, *The Causes of Evolution*, was published.[29] In it, Haldane summarized his mathematical models of natural selection and Mendelian genetics at a level that was accessible to the lay reader.[30] It was also in this book that he provided a hint of his mathematical model for kinship and the evolution of altruism—a model that he would lay out more explicitly in a 1955 article called "Population Genetics," published more than twenty years after *Causes*. Both were simple mathematical models, in that they were presented verbally, that is, without any variables or equations, but they mark a watershed in the study of altruism, evolution, and kinship. Finally, after Darwin, Huxley, Kropotkin, and Allee's discussions, and, to some extent experiments, on the topic of kinship and altruism, somebody—and in this case someone already recognized as a world-class scientist—had thought about the issue in mathematical terms.

It was not as if Haldane had never contemplated the importance of blood relatedness before penning *Causes of Evolution* in 1932. Indeed, many in his series of ten papers on natural selection dealt with the role of inbreeding, and hence interactions among kin.[31] More to the point, though, he explicitly recognized the role that protecting and aiding kin played in natural selection as early as 1928, when in an essay entitled "Darwinism Today," he noted, "Of two female deer, the one which habitually abandons its young on the approach of a beast of prey is likely to outlive the one which defends them; but as the latter will leave more offspring, her type survives, even if she loses her life."[32] In the deer case, we see hints of where Haldane is heading, for he always thought in mathematical terms, and so when he noted that a deer mother will leave more offspring even if she dies, he was already thinking about counting something that matters in the evolutionary process. It is unclear, though, what the "something" actually was. It may have genes coding for altruism, which would represent a major conceptual breakthrough, or it may have been how many "altruistic types" make it into the next generation, which is still a breakthrough, but not as fundamental as that of counting genes. Since Haldane often thought in terms of genes, it is certainly possible that he was looking at altruism from a gene-counting perspective—and from that point of view, it does not matter whether the genes coding for altruism reside in a female deer or in her kin. All that matters is that they make it into the next generation.

As his intellectual idol, Thomas Henry Huxley, did before him, Haldane made his position clear on any moral implications of the deer example. In his typical, straightforward manner, he noted, "Of course, the fact that nature favours altruistic conduct in certain cases does not mean that biological and moral values are in general the same. As Huxley pointed out long ago, this is

by no means the case, and an attempt to equate moral and biological values is a somewhat crude form of nature worship."[33]
From the moment Haldane began thinking about altruism and
kinship, he recognized that this was not a strictly academic
question, but one with moral overtones. He may have found this
particularly appealing, as he viewed the scientific method as the
best yet devised to tackle any question, including those relating
to morality.[34]

Although he held much in common with the scientific philosophy of Huxley, politically Haldane's views were more closely
aligned with Kropotkin. At first, he was not overtly leftist, even
quipping that his favorite Marxist was Groucho. But with the rise
of the Nazi Party in the 1930s, Haldane moved more to the left,
and over the next ten years he became one of the most outspoken
members of the Communist Party in England. He eventually
took a job as science correspondent for the Communist Newspaper the *Daily Worker* and published his first article, "What Makes
the Birthrate Fail," on December 9. 1937. He ended his stint at the
Daily Worker thirteen years later with a piece called "They Want
to Sterilize the Poor." As Haldane's biographer Ronald Clark
notes, "In between came nearly 350 articles, invariably laced with
propaganda, yet explaining the scientific facts with a sureness of
touch hardly equaled since the days of T. H. Huxley."[35]

In 1938, Haldane published his manifesto on the subject of
communism in a book entitled *The Marxist Philosophy and the Sciences*, a tribute to communism and especially Soviet communism. Haldane viewed Soviet communism as a political system
particularly open to radical experiments, and he was inherently
drawn to such a philosophy. Yet, despite being both one of the
leading communists and one of the leading scientists in England, Haldane's work on evolutionary theory—and in particular his work on kinship and evolution—show no clear signs of
being affected by his political leanings. Indeed, as we have already seen, his emphasis on natural selection as a force that
weeds out inferior forms was at odds with the Soviet approach
to Darwinism. As far as it is possible to judge such things, he
seems to have cordoned off the part of brain associated with
radical politics from that associated with mathematical models
of evolutionary change in a way that Kropotkin could never
have imagined.

Haldane did not formalize his ideas on kinship and altruism until he laid them out, albeit briefly, in *Causes* and then later in "Population Genetics."[36] He intended *Causes* to be a general manifesto on the power of natural selection. The first page of the book went right to the heart of the matter, arguing ". . . a generation ago most people who believed in evolution held that it had been largely due to natural selection. Nowadays, a certain number of believers in evolution do not regard natural selection as a cause of it, but I think, that in general the two beliefs still go together."[37] Haldane clearly accomplished his mission; *Causes of Evolution* is required reading, even today, for all evolutionary biologists. In the course of a general argument about the power of natural selection, he took up the question of kinship and altruism.

Haldane addressed kinship and altruism in *Causes* while trying to make sense of genes that lower the fitness of the individual in which they reside but increase the fitness of others in their society. Such genes, he understood, are often doomed to be "extinguished by natural selection in large societies." For T. H. Huxley, and for Haldane, too, there was an exception to this general rule, and once again it occurred within the realm of family life. Extending his initial discussion of the subject to humans, he noted, "I doubt if man contains many genes making for altruism of a general kind, though we probably do possess an innate predisposition for family life. . . . For insofar as it makes for the survival of one's descendants and near relations, altruistic behavior is a kind of Darwinian fitness, and may be expected to spread as the result of natural selection."[38] Here, Haldane moves not only from the deer case to the case of humans, but from focusing on the parent-offspring relationship (as in the deer example) to the more general case of "near relations"—that is, other blood kin.

In the appendix to *Causes*, where he could be a bit more technical, Haldane phrases this slightly differently, arguing that altruism can spread in small groups if "the genes determining it are borne by a group of related individuals whose chances of leaving offspring are increased by the presence of these genes in an individual member of the group whose own private viability they lower."[39] In both cases, the message is clear: if a gene coding for altruism benefits blood relatives, it may spread through natural selection.[40] If the relatives are direct offspring, the process is more powerful, but it works even if blood kin are more distantly related.

Haldane deferred a more precise description of how kinship affects altruism to his 1955 "Population Genetics" article, where he asks the reader to imagine the following scenario in a small population:

> Let us suppose that we carry a rare gene that affects your behavior so that you jump into a flooded river and save a child, but you have one chance in ten of being drowned, while I do not possess the gene, and stand on the bank and watch the child drown. If the child's your own child or your brother or sister, there is an even chance that the child will also have this gene, so five genes will be saved in children for one lost in an adult. If you save a grandchild or a nephew, the advantage is only two and a half to one. If you only save a first cousin, the effect is very slight. If you try to save your first cousin once removed the population is more likely to *lose this valuable gene* than to gain it. . . . It is clear that genes making for conduct of this kind would only have a chance of spreading in rather small populations when most of the children were fairly near relatives of the man who risked his life.[41]

This passage is a precursor to the modern evolutionary theory of kinship and altruism, for here Haldane provides some explicit gene counting. If you have the gene for altruism, there is a 50 percent chance that your child will inherit it from you, which is to say that the genetic relatedness between parent and offspring is 0.5 (fig. 4.1). In his scenario, if you have the gene for altruism, then for every ten of your children you save, you are saving, on average, five copies of the gene in question. If the parent in question has a one-in-ten chance of drowning per rescue, then over ten such rescues, on average, a single copy of the altruism gene (10×0.1) will be lost. On an "altruism gene spreadsheet," that equals five pluses and one minus, for a net gain of four.[42] Haldane did not present such an accounting to the reader, but it would have been next to impossible to arrive at his conclusions without implicitly thinking along these lines.

If grandchildren are in need of rescue, the net benefit received by the altruist is cut in half, because the chances that your

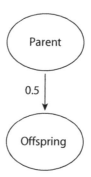

Figure 4.1. Genetic relatedness between parent and offspring. The r value between parent and offspring is 0.5. From the parental perspective, the probability of passing a specific allele on to its offspring is 0.5, as gametes (eggs and sperm) contain only a single copy of each gene a parent possesses, while every other cell in a parent's body contains two copies of every gene. From the offspring's perspective, its genetic relatedness to each parent is also 0.5, as it receives one copy of each gene from each parent.

grandchild has your gene for altruism is one-half times one-half, or one-quarter (fig. 4.2). On the gene spreadsheet, however, we still have a plus 2.5 (10×0.25) in the benefit column and a 1 in the cost column, and so the altruism gene should still evolve, as 2.5 minus 1 is still greater than 0. When we get to first cousins once removed, the math is trickier, but you share one-sixteenth of your genes with them. Here, the net benefit ($10 \times 0.0625 = 0.625$, or five-eighths) does not outweigh the cost (1), and hence selection should not favor such a gene.

Somewhat remarkably given his mathematical tendencies, Haldane did not derive a general equation that captures these costs and benefits and their relation to kinship. And, in particular, he made no attempt to understand how natural selection might act to maximize rules distributing altruistic acts among kin. If he had, the modern theory of kinship would likely be attributed to him. As it was, some time would pass before such equations would be developed, and while Haldane is given the credit for a very interesting thought experiment, the credit generally stops there. Indeed, in all the many written tributes to his work, his ideas on kinship and altruism are almost never mentioned.

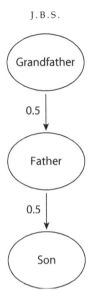

Figure 4.2. Genetic relatedness between grandfather and grandson. The probability that grandfather passes a specific allele on to father is 0.5, as is the probability that father passes that allele to son. As such, the probability that grandfather and grandson share an allele is the product of these probabilities, 0.25.

Haldane, like Darwin and Allee, took special note of social insects in developing his ideas on kinship and altruism, remarking "... the altruism of the social insects is more thoroughgoing. That is why moralists tell us to imitate them." J.B.S. was well aware of the high genetic relatedness shared by females in a hive of bees or wasps or in a nest of ants. Given the logical underpinnings of his "drowning" scenario, he realized that the high relatedness seen in female social insects would translate into natural selection favoring altruism, going as far as to say, "In the case of social insects there is no limit to the devotion and self-sacrifice which may be of biological advantage in a neuter. In a beehive the workers and young queens are samples of the same set of genotypes, so any form of behavior in the former ... which is advantageous to the hive will promote the survival of the latter, and thus tend to spread through the species."[43] In other words, if beehives are full of blood relatives, and neuters are not going to reproduce themselves, then natural selection should favor neuters expressing all sorts of altruism toward their reproductively active sister, the queen. Here, again, Haldane implicitly builds a model of genetic relatedness and altruism, but fails to construct an explicit mathematical model.

The modern-day, mathematically trained evolutionary biologist reading Haldane's *Causes* and his later "Population Genetics" is left somewhat in the lurch by the nonmathematical nature of the discussion on kinship and altruism. True, Haldane seems not only to have outlined a theory of kinship and altruism but to also have anticipated the large role that insects would play in testing any such models. Yet, in almost all his other works, J.B.S. translated his major ideas into mathematical equations—indeed, he ends *Causes* by telling the reader: "The permeation of biology by mathematics is only beginning, but unless the history of science is an inadequate guide, it will continue." And so we are left with a sense of abandonment on the question of altruism and kinship, where the mathematics that Haldane himself espoused is lacking.

The other two founding fathers of population genetics, Ronald Fisher and Sewall Wright, provided no more help in developing a formalized mathematical model of altruism and kinship than did Haldane. In *Causes*, Haldane noted early on, "I can write of natural selection with some authority because I am one of three people who know most about its mathematical theory."[44] He was certainly correct in this bit of self-aggrandizement, but that begs the question of what Fisher and Wright had to say on the matter of kinship and altruism.

Sir Ronald Aylmer Fisher, who was a colleague of Haldane's after J.B.S. moved to University College London in 1933, is credited not only with developing some of the foundational theories of modern statistical analysis but with further championing the power of natural selection as an evolutionary force. Fisher argued time and again that natural selection was a slow, methodical process. He thought in statistical terms and saw the need for general theorems that describe the evolutionary process, so that the study of natural selection "may be compared to the analytic treatment of the Theory of Gases, in which it is possible to make the most varied assumptions as to the accidental circumstance, and even the essential nature of the individual molecules, and yet to develop the general laws as to the behavior of gases,

TABLE 4.1
The Genetic Relatedness between Various Sets of Relatives

Generation	Brother	Half-brother	First cousin	Half-first cousin	Second cousin	Half-second cousin
Self	1/2	1/4	1/8	1/16	1/32	1/64
Father's	1/4	1/8	1/16	1/132	1/164	1/128
Grandfather's	1/8	1/16	1/32	1/64	1/128	1/256
Great-grandfather's	1/16	1/32	1/64	1/128	1/256	1/512
Great-great-grandfather's	1/32	1/64	1/128	1/256	1/512	1/1,024

Notes: For relatedness to your brother, half-brother, first cousin, and so forth, simply read across the "self" row. For relatedness to your father's brother (your uncle), your father's half-brother, and so forth, read across the second column. Do the same for relatedness to your grandfather's relatives (row 3), your great-grandfather's relatives (row 4), and your great-great-grandfather's relatives (row 5).

leaving but a few fundamental constants to be determined by experiment."[45] In a nutshell, Fisher, like Haldane, used both sophisticated mathematics and statistics to demonstrate that natural selection was a powerful force driving evolutionary change.

As Haldane had done in *Causes of Evolution*, Fisher synthesized his ideas into nonmathematical form for the general science reader in *The Genetical Theory of Natural Selection*,[46] a book that is more widely cited today than Haldane's *Causes*. Before this book, Fisher apparently had little to nothing to say about how kinship might affect the evolution of cooperation and altruism, although he was very interested in calculating degrees of genetic relatedness between individuals living in various types of populations.[47] For example, in one table found in his *Royal Society of Edinburgh* paper entitled "The Correlation between Relatives on the Supposition of Mendelian Inheritance," Fisher presented the degree of relatedness between thirty different sets of relatives (table 4.1).[48]

Although many complex mathematical equations preceded this table, and we know now that the entries in it are all correct, it is nearly impossible to follow just how Fisher came up with the values. At no point does he say anything like "to derive the genetic relatedness values I present in table 1, the reader need . . ." Even the most mathematically astute reader must take his word for the values he presented. The cryptic nature of this table was representative of Fisher's reputation in mathematical population genetics and appears to be the result of his early training. As a child, Fisher's eyes were extremely myopic, and he was forbidden to study under electric lights. To partly counter this disability, his tutor, W. N. Roseveare, instilled mathematics in him without any visual aids, which in essence forced Fisher to solve problems in his head. Indeed, many of his colleagues would express frustration that his mathematical proofs "contained intuitive leaps which were not obvious."[49]

Fisher's primary sortie into the relationship of altruism and genetic relatedness appears in a section of *Genetical Theory* entitled "The Evolution of Distastefulness." The biological issue at hand was a special form of defense used by certain insects. For Fisher, as for today's evolutionary biologists, the evolution of *most* defensive mechanisms in insects was easily explainable. For example, in the bombardier beetle, *Stenapitinus insignis*, individuals blast potential predators with a highly noxious spray.[50] Production of this spray is an act of engineering brilliance and a precursor to the techniques used in modern bomb making. The beetles have two glands, each of which has two separate compartments that store a set of different chemicals. The larger of these compartments contains hydroquanines and hydrogen peroxide, while the smaller compartment houses a variety of catalases and peroxidases. When threatened, bombardier beetles mix the contents of the two compartments, causing a chemical reaction that produces an acidic spray composed of *p*-benzoquinones. The heat produced by this chemical reaction causes an audible pop, and the acidic spray shoots out at a temperature of 212 degrees, Fahrenheit. Somehow, and it is not yet known how, the beetles are not injured by their own noxious sprays. What's more, they are quite adept at not only producing such antipredator sprays but also in selectively aiming them at predators.

Such "disagreeable" secretions as those produced in the bombardier beetle are easily explained by natural selection, as possession of these traits clearly benefits the possessor in its defense against predators. Fisher, however, puzzled over a related phenomenon, namely, "the process by which nauseous flavours as a means of defence, have been evolved."[51] And it was this issue—unpleasant flavors as a defense mechanism—that forced Fisher into directly tackling questions related to altruism and kinship. At first, he believed that for most adult insects, distastefulness only worked as a repellant *after* the predator began eating, and thus had already killed the prey. This sort of "defense" mechanism was difficult to understand as a product of natural selection, given that the victim was dead when the effects of a noxious taste affected the predator. But then Fisher learned from entomologist Edward Bagnall Poulton that he had been wrong in his initial assumption. Instead, he discovered that adult insects in distasteful species such as the monarch butterfly often have bodies tough enough to survive a predator's first bite. In this case, then, a predator could suffer the consequences of a noxious taste, stop, and, in so doing, the victim would survive. Standard natural selection models can be invoked to understand the evolution of this trait. This argument, however, could not explain distastefulness in insect larvae, who have much softer outer skeletons. It was when Fisher turned to distastefulness in insect larvae that he discovered the possible connection between kinship and altruism.

Larvae of many distasteful insect species live in groups, and with such gregarious larvae in mind, Fisher argued that the effect of distastefulness "will certainly be to give the increased protection especially to one particular group of larvae, probably brothers and sisters of the individual attacked."[52] If a predator responded to the noxious taste of a prey item, and in turn avoided ingesting the kin of the noxious (young) victim, then even if the initial victim died, natural selection at the level of the kin group could favor possession of such distastefulness in the prey species.

Fisher seemed on the verge of a mathematical model of kinship and altruism when he next noted that by saving the life of a nearby sibling, the deceased (distasteful) individual receives a genetic benefit because it is related to its sibling. He argued that when a predator abandons a site after ingesting a distasteful individual, *many* of that individual's sibs might be saved, and hence the genetic benefit to the deceased could be very substantial. Natural selection would then favor life-sacrificing altruistic traits such as distastefulness. Fisher was clear that his hypothesis about kinship and altruism holds true even in cases in which larvae are surrounded by both siblings and unrelated individuals. He recognized, however, that in such mixed-family instances, "The selective effect . . . will be diluted."[53] That is, as the average relatedness of individuals in a group falls, as is the case when unrelated insect larvae mix together, so to do the putative effects of kinship, and hence altruism is less likely to evolve.

The analogy between Darwin's difficulty with a bee's stinger being ripped from its body to serve its function and Fisher's problem with distastefulness in insects is striking. In both cases, a trait exists that seems to achieve the desired result only after the death of the individual in which it resides. Similarly, both Darwin and Fisher explained the evolution of such altruistic traits—those that benefit the group but pose a huge cost to the individual—as the result of selection acting at the level of the blood family. However, Fisher differed from Darwin in that he almost always went on to formalize his ideas as mathematical equations. Yet like Haldane, when it came to kinship and altruism, he seems to have abandoned this approach, and he built no mathematical model of altruism and kinship.

The logic that Fisher developed for understanding the evolution of distastefulness in insects caused him to speculate on the role of kinship in understanding the evolution of self-sacrificial traits in our own species. In particular, he was interested in the evolution of heroism in tribal societies. Heroism, by definition, entails some sort of risk or cost to a hero. What then favors the evolution of such a trait? While the situation was a bit more complex in humans, Fisher just adapted his distastefulness hypothesis to the problem of human heroism, and argued that such a trait could evolve "by the advantage which it confers, by repute and prestige, upon the *kindred* of the hero."[54] In tribal societies, Fisher argued, the benefits of belonging to a family that included a hero—power, prestige, access to resources and the like—compensated for the loss of a heroic family member. Even Fisher's foray into human altruism and kinship, however, was insufficient to foster the development of a mathematical model linking these phenomena.

Sewall Wright, W. C. Allee's colleague at the University of Chicago, was the third of the founding fathers of mathematical population genetics. Like Haldane and Fisher, Wright, too, was interested in the genetic relationship between relatives and published such papers as "The Biometric Relation between Parent and Offspring."[55] His most important contribution on the subject of genetic relatedness came in his 1922 paper entitled "Coefficients of Inbreeding and Relationship." Here he introduced a variable he called the "coefficient of relationship" (denoted as r).[56] Imagine two individuals, mother (1) and son (2). As we have seen already, their relatedness is 0.5. One way to denote this is to say that a son's relatedness to his mother is simply 0.5. Wright showed that the relationship between an individual and any one of its direct ancestors is 0.5^n, where n is the number of generations that separate our hypothetical individual from its hypothetical ancestor; one generation for mother ($r = 0.5^1 = 0.5$), two generations for grandmother ($r = 0.5^2 = 0.25$), three generations for great-grandmother ($r = 0.5^3 = 0.125$), and so on.

Once Wright calculated the degree of relatedness between ancestor and direct descendant, he moved on to the more general case of the coefficient of relationship between any two relatives, not just direct ancestors and descendants. Imagine that we have two relatives, labeled A and B, and that they share recent common ancestors, C1, C2. For example, if A and B are siblings, C1 and C2 would be their parents. Or A and B could be maternal first cousins, in which case the most recent common ancestor they shared would be a maternal grandmother and maternal grandfather.

Let's consider the case of the genetic relatedness between maternal cousins in more depth. Wright demonstrated that to calculate the genetic relatedness here, denoted r_{AB}, we begin by calculating the probability that maternal cousins both possess a gene—for example, gene 1—that they inherited from their maternal grandmother. As we have seen, the chance that individual A and its grandmother both have gene 1 is simply A's genetic relatedness to its grandmother, which is equal to $0.5^2 = 0.25$. The same reasoning demonstrates that the probability that B and its grandmother both have gene 1 is again 0.25. The chance that A and B both received gene 1 via their maternal grandmother is the product of these terms, or $0.25 \times 0.25 = 0.0625$. This same procedure is employed to calculate the probability that A and B both received gene 1 via their maternal grandfather, which again is equal to $0.25 \times 0.25 = 0.0625$. The probability that A and B share genes from *either* of their maternal grandparents—that is r_{AB}—is the sum of the two values, or 0.125 (that is, one-eighth). In other words, cousins have an r value of one-eighth, which translates into their having a one-in-eight chance of sharing a gene that they inherited through their maternal grandparents. The power of Wright's model is that it can be used to calculate the degree of genetic relatedness between any set of relatives in the same manner in which we used it for cousins.

Wright's calculation represented the first time that degrees of relatedness between relatives had been put in the form of easy-to-use equations, and his variable r would play a critical role in the modern theory of altruism and relatedness developed in the early 1960s. Yet he never attempted to quantify how r might affect the evolution of cooperation and altruism. What makes this especially odd in Wright's case is that he did build the group selection model of cooperation and altruism described in chapter 3, but he never directly linked that model to his work on genetic relatedness.

Why is it then that none of the founders of mathematical population genetics—men who made their living creating equations—built a model linking kinship and altruism? It may simply have been that Haldane, Fisher, and Wright thought the subject matter uninteresting. If this were the case, though, it is hard to explain why Haldane and Fisher would devote any space to the subject in *Causes* and *Genetical Theory*, respectively. Each viewed his book as *the* summary of his general views on evolution, and both omitted discussions of many of the issues they had touched on in earlier papers in order to provide the reader with a general theory. Given such self-imposed constraints on space, it would be hard to argue that, at least in the cases of Haldane and Fisher, the relationship between kinship and altruism was simply viewed as uninteresting. It is, of course, certainly possible that whereas the fathers of modern population genetics themselves found the question of kinship and altruism interesting enough to write about, they may have believed that others would have little interest in such matters, as work on the evolution of behavior had not yet come into bloom. Under such conditions, they may have felt obliged to touch on the subject of altruism and kinship, but go no further.[57]

An alternative explanation is mathematical arrogance on the part of Haldane, Fisher, and Wright. They may all have believed that a mathematical theory of kinship and altruism was unnecessary, because anyone with any mathematical know-how should easily be able to use their published material to see what such a model would look like. Haldane, for example, may simply have assumed that after he discussed the "drowning child" scenario with its set of costs and benefits and its description of genetic relatedness, the general mathematical case was made and any formal model would be trivial. Fisher, too, may have believed the same thing regarding his example of distastefulness. They all may have decided that there was no need to bother deriving a few equations when it was obvious (to them) what these equations might look like. In Wright's case, he may just have felt that it was clear to the reader that the small populations he modeled *could* easily be groups of kin, and hence his models already subsumed the notion of kin groups, making additional models superfluous.

This arrogance hypothesis, however, does not stand up to closer scrutiny. To begin with, in many ways Haldane, Fisher, and Wright were in fierce competition to produce the first comprehensive family of mathematical models of evolution by natural selection. If any one of them had thought of an explicit model of kinship and altruism, he would likely have published it, even if he was convinced that it would not be an earth-shattering find.

The arrogance hypothesis also fails to take into account the fact that all three of the founders of mathematical population genetics had developed models that examined genetic relatedness per se and knew that such models were not simple to construct. While it seems fairly easy to determine a mother and child's genetic relatedness of 0.5, Haldane, Wright, and Fisher all knew that such calculations could quickly get very complex, depending on what sorts of assumptions were made about such things as inbreeding and population structure. Adding inbreeding into their models, for example, complicates things enormously, as it requires recalculating the genetic relatedness now that blood relatives mate with each other.[58] Thus, they would not necessarily have considered it easy to build a model of kinship and altruism. Indeed, as we shall see when we examine the modern theory of kinship, while the kernel of modern theory can be reduced to a single, simple equation, a detailed mathematical analysis underlies that equation. And without an equation with specific variables that can be measured and put to the test, it is hard to make any specific predictions about the manner in which kin-based altruism should manifest itself in nature. Without a mathematical model, we can safely say that kin-based altruism will often be selected, but we cannot say much more than that.

During a visit to Cornell University, when I raised the question of why Haldane, Fisher, and Wright never came up with a model of kinship, another hypothesis surfaced: since none of the three founders of population genetics spent any amount of time in the field observing altruism and the role kinship plays in promoting it, perhaps they all lacked the natural history experience needed to properly model the phenomenon. Thomas Seeley, who studies honeybee ecology, evolution, and behavior at Cornell argues, "They weren't naturalists; they related to the world through mathematics. They weren't looking at behavior; they didn't see it as an important phenomenon."[59] Kern Reeve, also at Cornell, speculates that "all of the best models are born in the mind of somebody who has been in the field."[60] As Haldane's former student John Maynard Smith notes, Haldane, Fisher and Wright were all enamored with mathematical models, and though they all chose to tackle what they believed to be important questions, they were as far from naturalists as one can be.[61]

Alan Grafen has suggested a particularly intriguing hypothesis for why Haldane, Fisher, and Wright all stopped short of developing a full-blown model of kinship and altruism.[62] He suggests that the economic-based approach to evolution and behavior that would come to dominate work in this area in later decades was not yet in place. For example, Haldane and Wright did not conceptualize natural selection as an optimization process (though Fisher did), and hence would not have considered developing a model for how selection might optimize any rules with respect to kin-based altruism. In addition, for the most part, the three did not think about the costs and benefits of altruism as variables—things that can take on all sorts of values—and without that recognition, a mathematical model of kinship and altruism would be hard to envision.

In the end, we can only speculate as to why Haldane, Fisher, and Wright did not build explicit models of kinship and altruism. What is known, however, is that when William D. Hamilton, a born naturalist, published his seminal models of altruism and kinship in 1963 and 1964 (the year Haldane died), he made his debt to Fisher, Wright, and Haldane—the grand triumvirate, as he called them—very clear.[63]

Hamilton's Rule

Believing in the explanatory power of evolution by
natural selection is like a migraine. . . . The majority of
humanity seem to have difficulty in accepting that the
'oddness' of such a believer can be real—that is, simply
an oddness and nothing else. . . . As the migraine sufferer
is suspected of malingering . . . so the evolutionist is
always suspected of covert agendas unconnected with
reality or the search for truth. . . . Whatever the nature of
the ailment, I certainly caught or inherited it early
compared to most, and it burned me badly.
—*William D. Hamilton*[1]

WILLIAM DAVID HAMILTON grew up in a small British country
cottage outside of the small town of Sevenoaks, Kent. The cottage itself was called "Oaklea," after Hamilton's father's old
home in New Zealand. Oaklea sat atop a hill named Badger's
Mount—a hill that served as a base for the endless natural history romps that he took as a child. Born on August 1, 1936, Bill
Hamilton was the second oldest of Archibald Hamilton and Bettina Collier's six children. Archibald was an engineer who was
best known for his development of the Callender-Hamilton
bridge design.[2] The income from this innovation, though mod-
est, supported the household and allowed Bill's mother, one of
only a handful of New Zealand women to be granted a medical
degree, to give up her medical practice to raise her children.

Bill Hamilton was an avid naturalist and led the life of an ex-
plorer and outdoorsman from age ten onward, developing an
early passion for studying insects. He never had a bed in Oaklea, let alone a bedroom, and during the winter he spent his
nights on an army cot in a corner of the dining room. He preferred the warmer months of the year when he could sleep in
one of the many sheds surrounding the cottage where, according to his sister Mary Bliss, "it was quiet and much roomier."
She recalls Bill's entomological sorties taking "every free minute"

of his spare time.[3] When he was not in school, Hamilton could almost always be found on his bicycle heading for the fields and meadows that surrounded Oaklea, often in hot pursuit of some insect, most often a butterfly. He was keen on displaying his most recent catches to any sibling or parent willing to have a look. As he matured into his teenage years, however, he started to view insects not only as intrinsically beautiful but as a window into the laws of nature and a means by which he could try to decipher patterns and processes. This was in part owing to an inherent curiosity and in part a function of receiving a copy of E. B. Ford's book *Butterflies* as a birthday present—a book that "introduced the 12-year old to genetics, to a scientific sensibility that looked down on 'mere collecting,' to mathematical biology in the shape of Mendelian segregation ratios, and to the modern study of evolution."[4]

Archibald and Bettina Hamilton instilled an inquisitiveness in their children that produced independent and bold—sometimes too bold—individuals. Twelve-year-old Bill, for example, who was always experimenting with whatever materials were in the vicinity, felt comfortable tinkering with the bomb makings that his father kept in one of the Oaklea sheds. He pounded the ingredients into a brass cartridge, which exploded, blew off parts of three of his fingers, and imbedded brass fragments in his eye and chest. His sister came home that day to find her "father mopping blood off the back steps." Hamilton was brought to the emergency room of the King's College hospital, where he underwent a very dangerous operation on his chest cavity.[5]

Soon after he recovered from his operation, Bettina taught him Darwin's theory of evolution by natural selection. His introduction to natural selection was all the easier, as Oaklea was only a four-mile walk from Darwin's Down House—a walk young Bill and his mother took many times. Ford's book and Darwin's ideas had a profound effect on the boy. All of a sudden, the endless observations that he had made on insects began to make sense. "I never looked back," Hamilton recalled. "I was certain that this [natural selection] was the key to unlock a very wide variety of those patterns of nature that had fascinated me."[6]

During his teenage years, Hamilton's interest in evolution grew. Behind a door, in a corridor adjacent to the kitchen at Oaklea, lies the remains of a photographic darkroom where he conducted his first genetics experiment. He began by lining up his five siblings and taking their picture. Then he used this photo to get a measure of each sibling's height. Lastly, he examined whether the distribution of heights matched the normal distribution that Charles Darwin's cousin, Francis Galton, had predicted for complex traits like height.[7] Hamilton knew full well that five different-aged individuals would hardly suffice to test any of Galton's ideas on the normal distribution, but it was the best he could do. In any case, this makeshift experiment nicely depicts the budding interest of an evolutionary geneticist.

Hamilton's interest in natural history and evolution developed into an interest in the evolution of behavior. In particular, his ideas about kinship, altruism, and evolution were set into motion early on. He recalls "being part of a fairly large family and realizing even as a child that I thought very differently about my family members than I did about everyone else." Reinforcing this early revelation about kinship and altruism were the behaviors of the honeybees that Bettina bred at Oaklea. These insects "primed me to see again," Hamilton noted, "that there was something very special about close relatives."[8] Determining what that "something very special" was would soon become a lifelong obsession.

If the bees that his mother kept at Oaklea were partly responsible for Hamilton's burgeoning interest in kinship and behavior, then his fascination with mathematics and the art of tinkering were instilled in him by his father. As an engineer, Archibald Hamilton was always building new devices, and he taught his son that making something work, not making it look good, was what really mattered. Years later, when he would be regarded by many as evolutionary biology's greatest theoretician, he harked back to his father's tinkering. When it came to his mathematical proofs, he noted, it was "not so much the eloquence of the model that I strove for as something that will work and illustrate a theme."[9] Although he eventually was considered the primary theoretician of his generation, Hamilton never thought of himself as particularly mathematical.

As a young man in his early twenties, he spent 1956 and 1957 serving his mandatory "National Service" as a recruiter in the British Army's Corps of Royal Engineers. The brass particles from the bomb-making incident were still lodged in his lungs, so

Hamilton was deemed unfit to be drafted for foreign service. His sister Mary recalls his spending "two of the most miserable years of his life sending men to places he dearly longed to visit himself."[10] Hamilton endured his National Service by dividing his free time between studying and collecting insects in and around the woods of Chatham and Gillingham, and by visiting his much beloved maternal great-aunt, Prudence Jackson, who lived nearby and shared his passion for natural history.[11]

While in National Service, Hamilton kept detailed journals. The most remarkable thing about the hundreds of entries he made is that they contain almost no mention of anything relating to his military service. Instead, these diaries are filled with natural history observations, such as "traversed down stream, across the road . . . hoping to find Scarlet Tigers [butterflies] but found none." It seemed clear that whatever path Bill Hamilton would take in life, insects would somehow or another play a significant role. The journals reveal a keen observer who took notes on the height and species of trees surrounding him, the relative frequency of orchids, and even the presence of a particular woodcock who appeared each evening "at about or just after sunset." Hinting at young Hamilton's artistic leanings are wonderful pictures of butterflies and plants, scattered among natural history observations like "the night was a very good one for moths."[12] His journals also show a man who thought scientifically, that is, in terms of hypotheses and tests. The entries described experiments—some of which worked better than others—on cross-fertilizing different species of plants and on raising butterfly larvae under different conditions.

After his National Service, Hamilton began his undergraduate education at Cambridge University. As a young man set on studying biology, and in particular, natural history, evolution, and genetics, he was disappointed by what he saw at one of England's most elite universities. Despite the fact that the modern synthesis in evolutionary biology was well under way, it appears not have infiltrated the walls of Cambridge. To his utter amazement, Hamilton found that "many Cambridge biologists seemed hardly to believe in evolution or at least seemed to be skeptical of the efficacy of natural selection."[13] If he was going to learn more about evolution and genetics, he would have to do it on his own. And so he did.

He spent endless hours scouring the libraries at Cambridge for papers and books on evolution. One day while studying at Cambridge's St. Johns Library, he came across Fisher's book *The Genetical Theory of Natural Selection*. Hamilton "immediately realized that this was the key to the understanding of evolution." He became a self-admitted "Fisher freak" and immersed himself in *Genetical Theory* to the exclusion of everything else, including his course work at Cambridge.[14] Looking back on his days at college, Hamilton claims that Fisher's book "weighed as of equal importance to the entire rest of my undergraduate Cambridge B.A."[15]

In *Genetical Theory*, Fisher championed the idea that natural selection worked almost exclusively at the level of the individual and acted to maximize individual fitness. Yet in the halls of Cambridge, Hamilton heard something very different. There, in the rare instances when it was discussed, lecturers implied that natural selection acted to favor traits such as altruism and cooperation, not because of their effect on *individuals*, but because they helped preserve the *species*. This, despite the fact that, to Hamilton's surprise and delight, Sir Ronald Fisher still held his Balfour Professorship in Genetics at Cambridge. Hamilton met with Fisher on occasion, although the subject of their conversations is not known.

The basic idea being taught at Cambridge was that altruism protected a species from extinction, and that its evolution was somehow tied to this phenomenon. Everything that Hamilton had read by Darwin and Fisher suggested to him that this view was misguided. It was this tension that first ignited his interest in developing a mathematical theory for the evolution of altruism. He was certain that the Cambridge professors were wrong about altruism being favored because it preserved the species, but he also knew that there were no detailed mathematical models of altruism and its effect on individuals. And so he started developing such a model.[16]

Soon after he became interested in modeling the evolution of altruism, Hamilton realized that this subject had implications for understanding human behavior. Fortunately, he was enrolled as a genetics major at Cambridge, and the degree program in genetics required that students take some elective courses in other departments. Hamilton thought that a course in social anthropology might satisfy both the department of genetics mandate as well as his own interest in human behavior. To learn more about what the course covered, he spoke to its instructor, Professor Edmund Leach.[17] Leach, and the social anthropologists he represented, believed that *all* human behavior was culturally derived and that genetics had no bearing on the behavior of our species. For Leach, the idea that something as special as altruism could be explained by natural selection was tantamount to anthropological heresy. Leach was no heretic, and when he realized that his prospective student was interested in evolution, genetics, and altruism, he became unreceptive to anything Hamilton had to say.[18]

The department of genetics's reaction to Hamilton's idea of taking a class in social anthropology was only slightly less vitriolic than that of the anthropologists. The geneticists at Cambridge told him they believed that social anthropology was more akin to poetry than to science, and that he would be wasting his time taking a class in it. Hamilton decided not to enroll in the course, but the "blank refusal of both sides even to discuss the issue" was the single most important reason he decided that he would not remain at Cambridge for his graduate work.[19] Indeed, this whole episode did more than sour Hamilton on Cambridge University; it caused him to reconsider a career in science. "I am beginning to find Cambridge intolerably oppressive," he wrote to his sister Mary. "I think I will give up the hope of making headway against all this," he continued, "and take up school teaching and do my research on my own—after all it involves hardly anything but reading."[20]

After graduating Cambridge University, Hamilton reconsidered graduate studies in genetics, evolution, and behavior. He adopted a dual course of action: he applied to graduate schools and, at the same time, searched for schoolteaching jobs and acceptance to a program where he could earn a Diploma of Education degree.[21] His hopes for a teaching degree came to a standstill, however, when his favored program—the School of Education at Moray House (Edinburgh)—refused to accept his Cambridge genetics degree as evidence that he was capable of teaching science in secondary school. Hamilton believed that the response from Moray House may have been similar to that which he had received from Edmund Leach, and that the people at Moray House "were so 'nurture'-inclined that the word 'genetics' simply terrorized."[22] Whatever the reason for the rejection at Moray House, from that point on, his search focused solely on finding a graduate program in genetics.

Hamilton's transition to graduate school was a bumpy one. One of his top choices for a graduate program, the Galton Laboratory of Genetics at University College London, initially showed little interest in having him pursue his study of the evolution of altruism there. Hamilton received a cool greeting when he met with Professor Lionel Penrose, the director of the Galton labs and a colleague of J.B.S. Haldane. Penrose, a classic geneticist, and a leader in the antieugenics movement, was completely removed from the growing school of biologists that were beginning to tackle the evolution of behavior. Like Edmund Leach at Cambridge, Penrose was appalled that Hamilton intended to study behavior, particularly altruistic behavior, from an evolutionary perspective. To him the idea smacked of eugenics, and he told Hamilton that he wanted no part of his work on altruism— indeed that there was no "altruism problem" to be solved. Penrose did, however, recognize the value of Hamilton's Cambridge pedigree in genetics, and he was admitted into the graduate program in the Galton Laboratory of Genetics at University College.

Using his knapsack as a makeshift desk, many of Hamilton's days in graduate school were spent in libraries. He was a solitary person by nature, but there was more to it than that. He was not particularly impressed with others in his graduate school cohort. "I am only just beginning to realise," he wrote, "how extraordinary it is for a graduate student to have any ideas about what he wants to do; normally, they are only too keen to take up any project which the professor offers them. This sort of thing fills me with dismay and makes me wonder what science is coming to."[23]

At the same time that he was a graduate student at University College, Hamilton was jointly enrolled in a graduate program at the London School of Economics (LSE). While interviewing for admission there, Hamilton met Norman Carrier, a human demographer. Carrier liked Hamilton's idea of studying the evolution of altruism and encouraged him to apply for a prestigious Leverhulme Studentship. Hamilton was awarded this fellowship and officially enrolled in the human demography program in the LSE's department of sociology. At LSE, he wanted only "access to libraries plus some small pittance of support; these together would give me the freedom to follow my puzzles." As he saw it, "two colleges were kindly offering this and that was enough"[24]—never mind that he was to be associated with the department of genetics in one and the department of sociology in the other.

Looking back on his graduate student days, Hamilton recalls, "Most of the time I was extremely lonely. Sometimes I came to dislike my bedsitting room so much that, when even late libraries like Senate House or Holborn Public closed, and I was still in a mood to continue work, rather than return to my room, I would go to Waterloo Station, where I continued reading or trying to write out a model on the benches among the waiting passengers in the main hall." The crowds, though full of complete strangers that Hamilton described as "alcoholics there sheltering or craving company like me . . . lovers parting . . . fractious children herded by tired mothers," muted his loneliness. Hamilton would also take his work to the benches at Chiswick House Garden or Kew Gardens, and at times become lost in the natural beauty that surrounded him. It was in these libraries, gardens, and train stations that Hamilton, alone, began to piece together his ideas on evolution, kinship, and altruism and to realize that the evolutionary biologist is both blessed and cursed with "a fourth intellectual pigment of the retina capable of raising into clear sight patterns of nature and of the human future that are denied the majority of his fellows."[25]

When Hamilton began examining the role that blood kinship plays in the evolution of altruism, he considered himself "a crank." "How could it be," he asked, "that respected academics around me, and many manifestly clever contemporary graduate students I talked to, would not see the interest in studying altruism along my lines unless it were true that my enterprise were bogus in some way obvious to all of them but not me?"[26] Indeed, most of the people that he talked with about the evolution of altruism refused to acknowledge that there was even a problem that needed to be solved.[27] Hamilton found that his colleagues thought that "what little was worth saying about it [kinship and the evolution of altruism] certainly had been said by J.B.S. Haldane."[28] Although thought-provoking, for Hamilton, Haldane's ideas hardly constituted a well-defined theory of kinship and altruism.

Unlike Kropotkin, Huxley, or Allee, Hamilton appears to have had no philosophical, political, or religious leanings that influenced his opinion about whether natural selection worked via kinship to produce altruism. He certainly did not have a sentimental picture of the world, as a place where altruism was pervasive; indeed, he held some controversial views on eugenics, on occasion seriously raising the issue of the use of infanticide for severely handicapped babies.[29] Instead, Hamilton developed his model of altruism for two reasons: because he thought that the study of evolution and altruism was a fundamentally fascinating and important topic that had not been sufficiently examined and because his observations of insects had demonstrated that kin-biased altruism was real.

Hamilton published his first scientific article, "The Evolution of Altruistic Behaviour," in the September–October 1963 issue of the *American Naturalist*.[30] He opened this three-page paper with a statement that later would become the war cry of sociobiology: "It is generally accepted that the behaviour characteristic of a species is just as much the product of evolution as the morphology." But, as Hamilton quickly added, there are some kinds of behavior that could not easily be explained by classic evolutionary thinking, "in particular . . . any case where an animal behaves in such a way as to promote the advantages of other members of the species not its direct descendants at the expense of its own."[31] He quickly dismissed any arguments that altruism evolved to preserve entire species "as unsupported by mathematical models" and then laid out, for the first time, his own model of blood kinship and the evolution of altruism.

Hamilton asked the reader to imagine a pair of genes—gene G, which codes for altruism, and gene g, which does not. Because G codes for an act that entails a cost to the actor but a benefit to others, while g codes for no such action, standard models of natural selection at the level of the individual—the sorts of models that Fisher had so eloquently developed in his book and in his "fundamental theorem of natural selection"—failed to account for the evolution of G. Under standard models, G is always at a selective disadvantage compared with g and hence should never increase in frequency. But, Hamilton argued, if the effects of kinship were added to the standard model—creating a more "inclusive" model—altruism could evolve.

To build his more inclusive model, Hamilton used Sewall Wright's "coefficient of relationship," r, as his measure of genetic relatedness.[32] The power of Wright's coefficient was that it was a continuous variable that ranged from 0 to 1. No longer would it be necessary to speak about the evolution of altruism between any specific set of kin (parents and offspring, siblings, etc.). With r as a variable, any degree of relatedness—including a genetic relatedness of zero—would be covered. Next, Hamilton added in the costs and benefits of altruism to his model. To see how this works, think of the benefit that a recipient of an altruistic act obtains as b, and the cost paid by an altruist as the variable c. For example, imagine an altruist who brings food back to its baby brothers and sisters. In such a case, the benefit might be an extra chick surviving in the nest of an altruistic bird, while the cost might entail an increased risk of death for the altruist.[33]

The power of creating two variables to cover the cost and benefit of altruism lay not in the way that these variables applied to any specific case (like the chick example just given). Instead, the importance of adding b and c to his model was that it allowed Hamilton to take an economic approach to how natural selection might maximize fitness and still allow for the evolution of altruism. In his graduate days at the London School of *Economics*, Hamilton must have been exposed often to this sort of cost-benefit optimization analysis. But applying it to an evolutionary problem—that of kinship and altruism—was a eureka moment for both Hamilton and the field of evolution and behavior.[34]

In Hamilton's model, natural selection favors the gene for altruism whenever $r \times b > c$. This equation has become known as Hamilton's rule, and it consists of the following: If a gene for altruism is to evolve, then the cost (c) of altruism must somehow be balanced by compensating benefits to the altruist. In Hamilton's model, the cost is balanced by benefits (b) accrued by blood relatives of the altruist, because relatives *may* carry the gene for altruism as well. But relatives have only *some probability* (r) of carrying the gene in question, and so the benefits received must be devalued by that probability. Phrased in the cold language of natural selection, Hamilton's rule recognizes that a gene for altruism can spread if it helps copies of itself residing in blood kin. The engine of goodness, it suggests, lies in the family unit.

The publication of Hamilton's 1963 paper, and the creation of a simple "inclusive fitness" formula, marked a watershed in the study of evolution and behavior. A hundred-year debate that included not just science, but politics, religion, and mental illness was at last resolved. In principle, scientists could now go out and measure relatedness, costs, and benefits and test a model of kinship and the evolution of altruism. The debate shifted from the question "Do kin help one another, and is there something special about family?" to the much more interesting "Under what precise conditions should blood kin help one another?" The former question engendered endless arguments; the latter promoted experimentation.

Part of the power of Hamilton's equation—$r \times b > c$—stems from the fact that it incorporated so many of the concepts that Allee, Haldane, Fisher, and Wright had discussed but not explicitly quantified. In fact, in a show of reverence to the founders of population genetics, there are only four references made in all of "The Evolution of Altruistic Behaviour": two to Haldane for his *New Biology* paper and his *Causes of Evolution*, one to Fisher's *Genetical Theory of Natural Selection*, and one to Wright for his work on r. Hamilton's rule gathered together not only Wright's genetic coefficient of relatedness, but Allee's emphasis on ecology. For the benefits and costs in Hamilton's equation only made sense when one understood the ecology of the organism being studied. For example, the effort that it takes a bird to help raise a sibling, or the cost of predation when doing so, can only be conceptualized in terms of the ecology of the bird species in question. And, of course, Hamilton's equation—which quantified the ideas that Fisher and Haldane had only talked about— was all about the role of kinship in the *evolution of altruism*, which would have pleased Darwin.

Hamilton's model also filled a psychological void for those studying evolution and behavior. Up until 1963, such scientists suffered from what one might call "physics envy," in that they saw mathematical equations as tools that not only promote hypothesis generation and hypothesis testing, but as objects that garner respect from other scientists. Very few such equations existed in the area of evolution and behavior before 1963, and those that did did not pack nearly the punch of Hamilton's equation.

With Hamilton's rule in hand, it became possible to quantify the ideas that Fisher and Haldane laid out in their books. If, as in Fisher's case of the evolution of insect distastefulness, altruism occurs between siblings, where $r = 0.5$, then the benefit that an individual receives for saving *one* of its siblings from a predator must be more than twice the cost that the altruist pays. If, as in J.B.S. Haldane's "drowning grandchild" example, $r = 0.25$, the benefit that the grandchild receives must be more than four times the cost the grandparent pays.[35] If we think in terms of the probability of drowning, then the odds that a grandchild is saved (does not drown) must be more than four times higher than the chances that the grandparent will drown trying to save his or her kin. Hamilton's rule not only explains altruism between siblings or between grandparents and grandchildren, but it explains altruism between *any set of relatives*. If we know the costs and benefits associated with an altruistic act, and we know the relatedness of the individuals involved, Hamilton's equation allows us to predict whether altruism will evolve for any value of r, b, and c. Indeed, Hamilton's rule allows us to make some very general predictions about the evolution of altruism. Because this rule is more easily satisfied when r is large, *the more related individuals are, the more likely altruism is to evolve*. In a similar vein, the greater the benefit/cost ratio for the altruist, the more likely altruism is to evolve.

Although Hamilton's rule was first formulated for the case in which an altruist helps a *single* relative, because of the economic nature of the model, it can easily be modified to examine a single altruistic act that has effects on *many* blood relatives simultaneously. In that case, altruism evolves when $(\sum_{1}^{n} r \times b) > c$. In this equation, $\sum_{1}^{n}$ sums benefits across *all* kin that are helped by the act of an altruist. If an altruistic act affects *two* siblings ($n = 2$), it will be favored as long as the benefit received by the recipient is simply greater than the cost to the altruist—exactly half the benefit that is necessary for altruism to evolve when an altruistic act benefits a *single* sib. To see this, substitute 0.5 for r and then multiply the left-hand side of the equation above by two (for the two siblings that are aided by the altruist). If the altruistic act aids four siblings ($n = 4$), it will be favored as long as the benefit received by each recipient is half the cost paid by the altruist, and so on.

"The Evolution of Altruistic Behaviour" was almost immediately eclipsed by another pair of Hamilton publications a year later, in 1964. Although the original paper introduced evolutionary biologists to what would be called Hamilton's rule, it was lacking in two important areas. First, the model presented was mathematically very simple, almost intuitive.[36] Evolutionary biologists, like most scientists, tend to prefer complicated models; the underlying assumption being that the more mathematically complicated the theory, the more important it must be. Hamilton himself "realized from common experience that university people sometimes don't react well to common sense, and in any case most of them listened to it harder if you first intimidate them with equations."[37] Second, the 1963 *American Naturalist* paper lacked empirical examples of Hamilton's rule at work in nature. In fact, it did not contain a single example of how to use $r \times b > c$ to explain a case of altruism in the wild.

Hamilton knew that there were deficits in his 1963 paper. In fact, long before he wrote it, he had been working on very long and detailed article entitled "The Genetical Evolution of Social Behaviour," which addressed both the theoretical and empirical problems mentioned above. The shorter, pithier *American Naturalist* article was published first in a move of desperation. As his days in graduate school neared an end, he believed he "urgently needed . . . to represent the fruits of the three years . . . doing research."[38] The *American Naturalist* paper served as a quick fix. The important details would come in "The Genetical Evolution of Social Behaviour." Hamilton submitted this second paper to the *Journal of Theoretical Biology* (*JTB*) in early 1963, after his *American Naturalist* article had been accepted for publication. One referee—Professor John Maynard Smith—demanded that major revisions be made before it was acceptable for publication. Maynard Smith also asked that Hamilton's manuscript be divided into a pair of companion papers. The revisions and split took Hamilton nine months to finish, but eventually his papers, entitled "The Genetical Evolution of Social Behaviour, I and II," were published in *JTB*.

Maynard Smith was a former student of J.B.S. Haldane, and he shared Haldane's tastes, including "a materialist philosophy, a liking for simple mathematical models, and a dislike of the British Upper class."[39] While he was serving as a reviewer for Hamilton's *JTB* article, an unusual series of events unfolded that caused a long-standing feud between John Maynard Smith and Bill Hamilton.

The trouble began on March 14, 1964, when Maynard Smith published a paper entitled "Group Selection and Kin Selection" in the highly respected journal *Nature*.[40] Maynard Smith's article was in part a reaction to a book that V. C. Wynne-Edwards had published in 1962. In *Animal Dispersion in Relation to Social Behavior*, Wynne-Edwards revived W. C. Allee's and Sewall Wright's ideas that social behavior, including altruism, had evolved as a result of natural selection favoring certain groups over others.[41] In "Group Selection and Kin Selection," Maynard Smith dismantled Wynne-Edwards's "group selection" hypothesis using a combination of verbal and mathematical arguments. In the course of these arguments, Maynard Smith suggested that a process he dubbed "kin selection," not group selection, was the primary force behind the evolution of altruism. By defining kin selection as "the evolution of characteristics which favour the survival of close relatives of the affected individual,"[42] Maynard Smith was creating a new catch-phrase for the very process that Hamilton had outlined in his *American Naturalist* paper; indeed, Maynard Smith cites this article in his own paper.

Maynard Smith's *Nature* paper was published before Hamilton's "Genetical Evolution of Social Behavior" appeared in *JTB*.[43] As *Nature* is one of the premier journals in science, papers published there tend to make a splash, and this was certainly the case for Maynard Smith's article on kin selection. Hamilton was infuriated that Maynard Smith would publish a high-profile piece on altruism, kinship, and genetics before his own 1964 magnus opus on the subject appeared in print. He was particularly angry because not only did Maynard Smith know about his forthcoming *JTB* paper, but he was the very person who slowed down its publication. Over the course of many years, Maynard Smith apologized numerous times for the manner in which events unfolded in 1963–64.[44] Despite the apologies, however, when Hamilton's close friend, Harvard professor Naomi Pierce, recalled this incident, she noted rather somberly, "I don't think Bill ever forgave Maynard Smith."[45]

By splitting "The Genetical Evolution of Social Behaviour" into two parts, Hamilton was able to explicitly develop a more detailed mathematical model (part 1), and to discuss empirical work relating to altruism and kinship (part 2). In the opening to part 1, Hamilton used a heuristic tool mastered by Darwin in *On the Origin of Species*. In preparing the reader for his radical theory on evolution by natural selection, Darwin chose to open by introducing a subject with which the reader was very familiar— artificial selection. Indeed, the first chapter of the *Origin* goes into considerable detail on artificial selection programs in pigeons. Darwin then demonstrated that the same forces that govern artificial selection govern what he called "natural selection." It was a brilliant strategy. Fair-minded readers were forced to admit that if they accepted the familiar process (artificial selection), there was no logical basis on which to deny Darwin's new process (natural selection).

In Hamilton's case, he began "The Genetical Evolution of Social Behaviour, Part I" with a brief discussion of a subject his readers knew well—parental care in animals. To explain the evolution of parental care, Hamilton argued that parents help their offspring because offspring and parents share an r value of 0.5. This sort of explanation, although not usually cast in explicit genetic terms, was implicit in most work on parental care and would have struck readers as more than reasonable. Immediately following his explanation of parental care, he noted "there is nothing special about the parent-offspring relationship except its close degree.... The full-sib relationship is just as close $[r = 0.5]$.... Similarly, the half-sib relationship is equivalent to that of grandparent and grandchild with the expectation of replica genes ... standing at one-quarter; and so on."[46] Because of this, Hamilton argued that kinship-based altruism need not be restricted to the case of parents and offspring; instead, what was needed was a continuous measure of relatedness (r). If readers found Hamilton's explanation of parental care reasonable, they were forced to admit that Hamilton's argument about dispensing aid to more distant relatives was equally sound.

After introducing the reader to kin-biased altruism, the primary function of part 1 of the *JTB* paper was to provide a detailed mathematical model for the evolution of altruism.[47] Hamilton began this process by taking a somewhat different approach to modeling altruism and kinship than what he had adopted in the *American Naturalist* paper. Rather than beginning by addressing whether a trait for altruism can spread as a function of kinship, Hamilton started the *JTB* model by expanding the classic definition of fitness. Before Hamilton, evolutionary biologists typically defined fitness as an individual's lifetime reproductive success. If you want to predict whether natural selection will favor gene *1a* over gene *1b*, you examine the effect of these different genes on fitness; whichever gene has the greatest positive effect will then be favored by selection. Hamilton expanded this view to take a more direct "gene's-eye perspective" on fitness—an approach championed a decade later by Richard Dawkins in *The Selfish Gene*. Rather than simply counting up the number of offspring an individual produces in his or her lifetime, Hamilton argued that from the perspective of a gene for some trait—altruism, for example—what really matters is how many copies of that gene make it into the next generation.

One way a gene gets more copies of itself into the next generation is by increasing classic fitness—that is, the number of offspring produced by the individual in which the gene resides. But, in his 1963 and 1964 papers, Hamilton shows that there are other ways that genes get copies of themselves into the next generation, and one such way is to aid copies of themselves that are residing in other individuals. Blood relatives, by definition, are likely to have similar genetic compositions, and so from the gene's perspective, you are likely to get more copies of yourself into the next generation if you aid blood relatives of the individual in which you reside. Hamilton's model expands the classic definition of fitness, then, by making it more "inclusive" in the sense of recognizing the various ways genes can get more copies of themselves into the next generation.

To do this, Hamilton first breaks down the classic definition of fitness into parts and then rebuilds a model of inclusive fitness from scratch. In essence, what he does is ask what the reproductive success of an individual would be if that individual were unaffected by anything in its social environment. The social environment is then added back into Hamilton's model by "augmenting by certain fractions of the quantities of harm and benefit which the individual himself causes to the fitnesses of his neighbours."[48] The fraction Hamilton refers to is measured by

Wright's r term, and the evolution of altruism per se is built in by the "benefit which the individual causes to the fitness of his neighbours."

To show all this mathematically, Hamilton takes the reader on a journey through pages upon pages of complicated algebraic equations, calculus, and statistics, in which he slowly substitutes back in the r, b, and c terms used in his 1963 *American Naturalist* paper. At the end of part 1 of "The Genetic Theory of Social Behavior," all the mathematics lead the reader straight back to $r \times b > c$—Hamilton's rule.[49]

Evolutionary biologists and animal behaviorists had never seen anything like what Hamilton had done, and so he felt obliged to open part 2 with a verbal summary of the mathematics in part 1. "In the hope that it may provide a useful summary," Hamilton wrote, "we therefore hazard the following generalized unrigourous statement of the main principle that has emerged from the model. The social behavior of a species evolves in such a way that in each distinct behavior-evoking situation the individual will seem to value his neighbors' fitness against his own according to the coefficients of relationship (r) appropriate to that situation."[50] Once a mathematical and a verbal description of his ideas were presented, Hamilton speculated on how these ideas might shed light on specific examples of altruism in nature.

He began his empirical overview of altruism (part 2) with a discussion of "thumping" behavior in rabbits. If a rabbit sees a predator, it often thumps its back legs and raises its tail. While raising its tail, the rabbit flashes a white underside that warns other rabbits in the vicinity that a predator has been sighted. Those rabbits who see the warning signal respond by heading for cover. But why should any rabbit thump in the first place? Flashing a white patch and making lots of noise thumping must make the thumping individual the most obvious thing for yards around. Why pay such a high cost to help others? Hamilton argued that the key to unlocking this riddle was that those rabbits warned of the impending danger were not just a random sample of the population, but the thumper's blood relatives. From the perspective of inclusive fitness, the cost to the individual thumper is exceeded by the benefits accrued by its blood kin.

Hamilton also used inclusive fitness models to understand two other common behaviors seen in group-living creatures—

mutual preening and grooming.[51] Grooming and preening remove parasites from body parts that the beneficiary of these acts cannot reach. He hypothesized that the cost of grooming another group member was so small that almost any degree of blood relatedness between donor and recipient would more than make up for it.[52]

If, as Hamilton's theory suggests, helping blood relatives increases an individual's inclusive fitness, then altruism toward kin should be particularly forthcoming from individuals in their postreproductive years. "The behaviour of a post-reproductive animal," Hamilton argued, "may be expected to be entirely altruistic,"[53] since aiding even distant relatives provides some inclusive benefits—the only fitness benefits *post*reproductive individuals can accrue. To test this idea, Hamilton turned to A. D. Blest's work on saturnid moths. Blest had studied cryptic and aposematic coloration in moths. Cryptically colored species of moths use their coloration to blend into their environment and make themselves *less* obvious to predators. Moths that rely on aposematic coloration use their colors to warn predators that they contain noxious substances and hence would make for a bad meal. Hamilton argued that inclusive fitness thinking makes very different predictions with respect to the postreproductive behavior of individuals from cryptic versus aposematic species.

In the case of cryptic coloration, if individuals live in the vicinity of kin, then, Hamilton argued, inclusive fitness theory predicts that "it is altruistic to die immediately after reproduction."[54] To see why, imagine a postreproductive cryptic moth. If such an individual is spotted and eaten by a predator, that predator is more likely to learn what *all* cryptic moths look like and is then more likely to eat the nearby kin of the deceased. In such a case, the postreproductive moth's inclusive fitness would have been higher if it had simply died after its last bout of reproduction, rather than lived and potentially drawn an experienced predator to the area containing its kin.

In the case of aposematic species, intense coloration is almost always associated with a noxious taste, and so inclusive fitness thinking led Hamilton to a very different conclusion regarding postreproductive behavior. A predator who eats a postreproductive individual in an aposematic species will be *less* likely to eat the deceased individual's relatives, since it will have learned that aposematically colored individuals taste terrible. A *postre-productive* individual in an aposematic species may raise its inclusive fitness by being eaten, and hence selection should favor life after reproduction. To Hamilton's delight, the postreproductive life of cryptic and aposematic species matched that predicted by inclusive fitness models—with postreproductive life spans significantly shorter in cryptically colored moth species.

In a section of part 2 entitled "A Hypothesis Concerning the Social Tendencies of the Hymenoptera," Hamilton argued that the odd genetic systems of the hymenopteran insects—ants, bees, and wasps—makes them predisposed to high degrees of altruism. Female ants, bees, and wasps follow the typical mammalian genetic trajectory in that they develop from eggs that are fertilized by sperm. As such, females have the typical two sets of chromosomes—one from mother, and one from father. Male hymenopterans, however, develop from unfertilized eggs, and so they inherit only one set of chromosomes, and that comes from their mother. Males, then, have no father. The technical term for this genetic system is haplodiploidy, which is derived from fusing the terms haploid (one set of chromosomes), and diploid (two sets of chromosomes). This bizarre genetic system can make figuring degrees of genetic relatedness in social insects tricky, but a number of important results emerge when such calculations are made.

In most vertebrate systems, siblings share an r value of 0.5, but because hymenopteran social insect sisters share all of their haploid father's genes and half of their diploid mother's genes, their r value is inflated to 0.75—exactly halfway between that of typical siblings and identical clones ($r = 1.0$). As a result, worker females are more related to their sisters ($r = 0.75$) than to their own daughters ($r = 0.5$). This asymmetry in relatedness favors the evolution of a sterile female worker caste that helps the queen produce more and more of its superrelated sisters. Another outcome of haplodiploid genetics is that hymenopteran females are less related to their brothers than in other species. From the perspective of female hymenopterans they share only one common ancestor (their mother), with their brother and so their r value (0.25) falls exactly between that of unrelated individuals ($r = 0$) and typical siblings ($r = 0.5$).

Hamilton's inclusive fitness theory provided a mathematical foundation for interpreting the connection between the asymmetries in hymenopteran relatedness and the well-known behavior of both worker females and male drones. It is the sister workers that show such a relatedness value with the queen who demonstrate the altruistic behavior that so defines the hymenopteran, and it is the males, with their low genetic relatedness to females, that are the lazy drones, concerned only with mating opportunities. In the light of inclusive fitness theory, the suicidal sting and the virtual sterility of all females but the queen begin to make evolutionary sense. Female workers may be sacrificing their reproductive opportunities and even their own lives, but in return, they provide benefits to a hive full of sisters who, as a result of haplodiploid genetics, are almost their clones.[55]

No section of Hamilton's 1964 papers would eventually garner as much attention as that devoted to kinship, altruism, and the social insects. And it is easy to see why, considering the role these insects played in the history of this subject, the plethora of studies done on them, and the fascination we have with their complex societies. Even before Hamilton's papers, scientists had pointed out that there were other things besides the high genetic relatedness between females that made social insects special, and hence may help explain altruism in this group. Unlike many other groups of insects, for example, in the hymenopterans, parent and offspring generations overlap, and surely that is necessary for the sort of altruism they display. And since Hamilton's papers, there have been any number of new models that try to explain the evolution of altruism in social insects—papers that discuss such things as "assured returns on fitness," "insurance" against disaster, "reproductive head starts," and "ecological selection."[56]

At the end of the day, however, all the work after 1964 either uses Hamilton's inclusive fitness model as a foundation and then expands on it, or fails to explain why it is always female hymenopterans that are always the altruists. While there are certainly still heated debates over whether the r, the b, or the c term in Hamilton's rule is really the most important in explaining hymenopteran altruism, most (but certainly not all) students of behavior and evolution would argue that they all, one way or another, lead us back to Hamilton's 1963 and 1964 work.

Bill Hamilton was on his way from Brazil to Britain when "The Genetical Evolution of Social Behaviour, Parts I and II" were published in the summer of 1964. During the eighteen months leading up to the publication of his *JTB* articles, Hamilton was out in the rain forests of Brazil doing what he loved to do best—study insects in the wild. Sometime in early 1963 he had written to Warwick Kerr about joining his research group in Brazil to study evolution in the social insects. Kerr responded favorably, and Hamilton quickly began a year-and-a-half expedition in there.

From the opening lines of the journals that Hamilton kept on his 1963–64 trip, the reader immediately sees a man more at home with insects than people—virtually each of the 402 pages of journal entries that he logged say something about either collecting, observing, marking, dissecting, or experimenting with wasps.[57] So great was his passion for Brazilian insects, that in a short 1991 paper called "My Intended Burial and Why," Hamilton made it clear to all around him that when he died, he wished to be laid out in the Brazilian rain forest so that he could be buried by the giant *Copropheanaeus* beetle, wherein "They will enter, will bury, will live on my flesh; and in the shape of their children and mine, I will escape death."[58] Though it is clear from Hamilton's latter writings that he spent some of his early time in Brazil revising his *JTB* papers, one would never know that from reading his journals, where hardly a mention of this paper, or his theory of altruism and kinship, can be found. When he was out in the field, Bill Hamilton's mind was focused entirely on insects.

The Price of Kinship

AFTER HE RETURNED from Brazil in 1964, Bill Hamilton worked as a researcher at Imperial College's field station in Berkshire. This facility, called Silwood Park, had a stellar reputation for insect ecology and population biology, but it housed very few evolutionary biologists. So, when a lectureship in genetics opened at the London campus of Imperial College, Hamilton applied for the job. He obtained the position "with ridiculous ease." English institutions of higher learning were expanding rapidly in the early 1960s, and, as Hamilton remembers it, "jobs for academics almost fell from the trees."[1]

Unable to afford a house on the meager salary that a lecturer at Imperial College received, Hamilton moved into a small flat owned by Blue Star Garages Limited, and looked "out on their emblem reared over a filling station"[2] every time he peered out his window. The location of his office, however, made up for the view from home, for it was situated smack in the center of an entomological golden triangle—the Royal Entomology Society was one hundred yards away at 41 Queens Gate, a few steps further down the road was the Commonwealth Institute of Entomology,[3] and, across the street lay the British Museum. For a man who preferred to spend time in the woods watching his beloved insects, it is hard to imagine an urban setting that Bill Hamilton would have appreciated more.

Hamilton kept busy at Imperial College, constantly sharpening his skills as a theoretician. Shortly after "The Genetical Theory of Social Behavior, Parts I and II" appeared in the *Journal of Theoretical Biology*, he published the first mathematical model for the evolution of aging, again in *JTB*. This model shifted the emphasis on senescence from a mechanistic-based approach that equated aging with "breaking down" to one in which natural selection continuously pushes deleterious traits back in development.[4] A year later he revolutionized the theory of sex ratios in a landmark paper in *Science.* Hamilton's mathematical model of sex ratios demonstrated that although a 1:1 (female:male) sex ratio was often favored by natural selection, there were conditions under which selection would favor the sex ratio in a population being heavily biased toward females.[5] Between 1964 and 1975, Hamilton also published a new model examining the evolution of group living,[6] as well as a number of papers and book chapters on kinship and altruism.[7]

In time, the depth and breadth of these papers would establish him as one of the leading evolutionary biologists of the twentieth century. Yet, despite the demonstration of his talent as a theoretician, the late 1960s and early 1970s were not all that Hamilton had hoped they would be. With some notable exceptions that we shall return to later, until the mid-1970s, not many people seemed to take notice of his kinship papers. One person who did notice, however, was an enigmatic, unknown genius named George Price.

George Price was four years old when his father died in 1926. His widowed mother—a former opera singer and actress—did her best to keep the family's lighting company afloat, but it was rough going.[8] From such humble beginnings, Price was eventually accepted into the University of Chicago, where he earned an undergraduate degree in chemistry (1943) and then a Ph.D. (1946) in the same subject.[9] In 1947, he married Julia Madigan, and the couple quickly had two daughters. The marriage, however, lasted just eight years, and eventually crumbled in 1955 when Julia's deep devotion to Christianity and George's passionate attachment to atheism created irreconcilable differences between them.

Price spent the years between 1946 and 1957 moving from one research position to another, including chemistry instructor at Harvard (1946–48), researcher at the Bell Telephone Laboratories (1948–50), and research associate in medicine at the University of Minnesota (1950–57). While at the Minnesota, he also worked part time as a science journalist, publishing a lengthy *Science* paper on the lack of concrete evidence for extrasensory perception.[10] Price's eclectic interests turned to evolutionary biology in 1966. Using the money he received from an insurance settlement for a botched thyroid operation, he bought a ticket on the *Queen Elizabeth* and sailed to England. His sole purpose for the trip was to visit the British Museum, the Senate House Library, the

Museum of Natural History, and the University of London to study evolutionary biology; indeed, on his résumé, Price listed his occupation for 1967 and 1968 as "reading and writing in London on evolutionary biology, while living on savings."[11]

Sometime in 1968, Price came across Hamilton's models on kinship and altruism and wrote to Hamilton requesting a reprint of this paper.[12] Subsequent to the reprint request, Price and Hamilton exchanged a few letters, but this correspondence came to a halt when Hamilton left for another long research trip to Brazil. Upon Hamilton's return, the letters between the two men were resumed. Soon, Price, like Kropotkin, Huxley, Allee, and Hamilton, before him, became almost obsessed with questions relating with altruism and its ties to kinship. After he read Hamilton's altruism and kinship papers, Price was depressed. He had hoped that all goodness was somehow exempt from scientific analysis, but Hamilton's models seemed to demonstrate otherwise. To be certain that Hamilton was correct, he intensively analyzed all the mathematics in the model and concluded that "apart from small discrepancies" Hamilton's model was indeed correct.[13] Not many people possessed the skills to deconstruct Hamilton's model of altruism and kinship and check for accuracy, but Price certainly did. More important, Price realized that Hamilton had actually *under*estimated the power of his own model. To his amazement, Price had discovered that in addition to explaining altruism, Hamilton's model could be used to study the evolution of spiteful behavior.

In evolutionary terms, a spiteful act lowers the fitness of the individual committing it, but decreases the fitness of the recipient to an even greater extent.[14] To examine the evolution of spite, Price began with Hamilton's model and then developed a totally new way to model evolutionary change—a model that relied on a mathematical technique called covariance analysis. Covariance is a measure of how two distinct entities change in relation to one another. Imagine that our two variables of interest are the score that a student obtains on a history test (H) and the number of hours that the student has studied (S) for the test. When high values of S are associated with high values of H, a positive covariance exists, and when high values of one variable are paired with low values of the other variable—for example, if we replace the number of hours a student has spent studying with the number of hours a student has spent at the local pub—a negative covariance exists.

Price applied the mathematics of covariance to evolutionary change. He examined the covariance between what allele (or gene variant) an individual possesses and how many children it produces. If certain alleles were consistently associated with the production of many offspring, such alleles should increase in frequency. To see this, consider a population of individuals living in small groups, and in which the average genetic relatedness in the population is X. What Price found was that when individuals were in groups with lots of blood kin—who, of course, were likely to contain the same alleles that they do—then a gene coding for altruism had a positive covariance with the number of offspring an individual produced. Altruism could then evolve. This result was not new (although the math Price used was) and served to verify what Hamilton had found in his 1963 and 1964 papers. What was new was Price's finding that if the average relatedness within groups is less than the average genetic relatedness within the population—that is, when individuals in groups are "negatively" related—then spiteful behavior can evolve. This can occur because when individuals engage in spiteful acts, they are differentially directing them toward others who are very unlikely to have the same genetic composition that they themselves possess—that is, toward their genetic competitors. And, although rare, instances of spite in animals have been recorded.[15]

Price wrote a manuscript summarizing the results of his model and submitted it to *Nature*. The paper was rejected, but the editor noted that *Nature* would consider a revised submission. Price was in no mood to waste time, and, rather than revise the original paper and send it back to *Nature*, he planned to submit it to the *Journal of Theoretical Biology*. Bill Hamilton, who by this time had become Price's good friend, thought this idea was a mistake. Instead, he proposed a plot to help land Price's paper on covariance, kinship, altruism, and spite in *Nature* after all. It went like this: First, Price was to revise his original submission by changing it into a very general manuscript on the power of covariance analysis. While Price revised his paper, Hamilton, who was intent on getting a paper on kinship and altruism into *Nature*, would write a short companion paper demonstrating that his 1963 and 1964 papers could be reformulated in the language of covariance. Under such a reformulation, if the conditions were just right, Hamilton's paper demonstrated that spite could evolve.

The plot called for Price to submit his revised paper to *Nature*; then, one week later, Hamilton was to submit his manuscript. Price's paper was rejected—in fact, it did not even make the first cut—and was sent back without review. Hamilton's article, on the other hand was quickly accepted. Hamilton knew that he already had a reputation as a theoretician, and that Price did not, and his plot anticipated *Nature's* response to both papers. Then, according to plan, once Hamilton's paper was accepted, he wrote to *Nature* and said that because he had made use of Price's "powerful new method," he could not publish his own results "until the method he [Price] used was published."[16] The editors at *Nature* reconsidered their decision, and published Price's paper along with Hamilton's.[17] Price's joining the corps of those obsessed with kinship and altruism had resulted in the development of a new and powerful method for evolutionary analysis, and two papers in *Nature* to prove it.

Two months before his covariance paper appeared in *Nature*, "fire-spitting atheist" George Price underwent a religious epiphany. "On June 7th [1970] I gave in," Price told friends, "and admitted that God existed."[18] He could see no explanation, save some sort of supernatural intervention, for all that had happened in his life over the past few years. In particular, in an irony that turns the debate between religion and evolution on its head, he believed that his findings on covariance, altruism, and kinship were the result of divine inspiration. As Hamilton remembered it, Price "believed that the discovery he had made in evolutionary theory was truly a miracle. . . . God had given him this insight where he had no reason to expect it. It was ludicrous, he [Price] told me that he, a person who never understood or used statistics and had hardly known previously a covariance from a coconut, could have discovered the simple formula that should prove to be the most transparent yet found to partition and interpret the working of natural selection."[19]

Now a convert, Price no doubt regretted that his atheism caused the breakup of his marriage fifteen years earlier—from his new perch, it looked as if his former wife, Julia, had been right all along. In any case, he accepted that he was in God's hands, to do with as He pleased. If his purpose on earth was to develop scientific models of evolution and social behavior, so be it. If God had further use for him, Price believed, the Almighty would see to it that he lived to do what needed to be done. He joined the All Souls Church at Langham and adopted the Good Samaritan as his role model.[20] He then proceeded to donate virtually all of his money to helping the poorest of the poor: at various times he lived on skid row as a squatter, at other times he slept on the floor at the Galton Labs (where he had been given some space to work on his evolutionary models).[21] Price now lived the life of the altruists that he had only modeled mathematically before.

Price's interest in evolution was not restricted to covariance, kinship, and altruism. He was also intrigued by the fact that animal fights so rarely involved lethal tactics. Why should that be? Why do animals not fight to the death—as gladiators do—when they contest resources that will increase their reproductive success? While the issue of mortal combat is clearly different from that which he tackled in his work on altruism, the two share some commonalities. In the first place, such battles, when they do occur, are rarely between kin; and second, prosocial behavior need not be restricted to taking an action, but may also be expressed by *not* acting. Refraining from fighting, in other words, is also a type of prosocial behavior.

Rejecting the "good for the species" argument that was prevalent in the 1960s, Price believed that the mathematical theory of games, rarely mentioned in the evolutionary literature of the period, might provide the answer to why deadly fights were so rare. Created by mathematical economists such as John von Neumann and Oskar Morgenstern, game theory is used when the "payoff" that an individual receives for some action depends not only on its behavior, but on the behavior of its partner (or opponent). Price realized two things about game theory, evolution, and aggression. First, as assumed by game theoretical models, when two individuals contest a resource, the payoffs associated with being aggressive or passive clearly depend on the behavior of both individuals. Second, the mathematical idea of a payoff mapped almost perfectly onto the evolutionary notion of fitness.

Price developed a game theory model of aggression and found that even though an individual using a "fight-to-the-death" strategy receives a high payoff when interacting with pacifists, if too many individuals adopt this approach, the costs of fighting mount quickly and the payoff plummets. Price submitted his model of game theory and aggression to *Nature* in July 1968. Seven months later, the editor wrote Price and informed him that if he shortened his paper in accordance with the wishes of one of the referees, he would it accept it for publication. Too busy to spend more time on this paper, Price never bothered to make the requested revisions. And that is where things stood—for a while.

In a remarkable parallel with the Hamilton–Maynard Smith story, John Maynard Smith was the referee who suggested that *Nature* accept a shortened version of Price's paper. Over time, he became enamored with Price's findings and began his own exploration of game theory and its possible uses in evolutionary biology. After reading what he could on the theory of games, Maynard Smith developed his own game theoretical model of fighting. He knew that before he could publish his model, he needed to locate Price and request permission to cite his unpublished manuscript. Given that Price was often living as a squatter, this was no easy task, but, Maynard Smith recalls, "with *Nature*'s help, I contacted him."[22] Price knew of the Hamilton–Maynard Smith controversy and was reluctant to grant his permission. This hesitation disappeared after Price met Maynard Smith in person, and Maynard Smith offered to make him a coauthor on his paper—an offer Price was happy to accept. Their model was published as the cover story in the November 2, 1973, issue of *Nature*. Price was thrilled, and told Maynard Smith that this was "the happiest and best outcome of refereeing that I've ever had."[23]

Price would make of use the *Nature* paper in unexpected ways. When it was published, Price was living from hand to mouth. Part of his self-appointed religious mission of altruism was to help the victims of spousal abuse. One day a "belligerent alcoholic," who was furious that Price was protecting his abused wife, expressed his thoughts on this matter by urinating on the steps of the genetics building in which Price did his work, as well as harassing nearby students with a barrage of obscenities. Price took all this in stride, writing to his daughter Annamarie, "I expect that one cover-illustrated lead article in *Nature* compensates for one urination at the front entrance of the building."[24]

In December 1974, George Price spent a relaxing week at the Hamilton home. Just prior to this visit he had written to his daughter Kathleen that he was "heading back up," and had started to amass some personal belongings. When he left the Hamiltons on December 19, it was with the understanding that he would return shortly after New Year's Day. Sometime after he departed, Price's spirits plummeted. On January 6, 1975, he committed suicide by slitting his throat.[25] One of the few pieces of useful information the police found in Price's decrepit flat were some letters he had exchanged with Bill Hamilton. The letters contained Hamilton's address, and he was called to testify at the inquisition held after Price's death and also charged with cleaning out the flat. "As I tidied what was worth taking into a suitcase," Hamilton noted, "his dried blood crackled on the linoleum under my shoes; a basically tidy man, he had chosen to die on the open floor, not on his bed." Price's funeral was held in Camden Town. In a final scene that captured his life, the funeral was attended by only eight individuals, including Bill Hamilton, John Maynard Smith, and "two or three red faces framed in shaggy hair, the hair in turn touching the shoulders of old dark coats."[26]

Spreading the Word

THOUGH HIS WORK had a profound impact on people like George Price, Bill Hamilton's ideas on blood kinship and altruism would take time to spread to other scientists and to the lay public. Indeed, most people knew nothing of Hamilton's models on altruism and kinship ten years after they had appeared. This all changed, however, with the publication of two of the most popular evolutionary biology books ever written: Richard Dawkins's *The Selfish Gene* (1976), and E. O. Wilson's *Sociobiology* (1975).[1]

By the time that Oxford University Professor Richard Dawkins published his book *The Selfish Gene*, the study of behavior and evolution was well established; indeed, Lorenz, Tinbergen, and von Frish had received the Nobel Prize for founding the field of animal behavior (technically known as ethology) just three years earlier. Dawkins was trained as an ethologist at Oxford under the tutelage of Tinbergen himself and was an exception to the rule, in that he had taught Hamilton's ideas on kin selection as early as 1966.[2]

Following in the footsteps of Hamilton, in *The Selfish Gene*, Dawkins argued that to understand how natural selection operates, we need to adopt a gene's-eye view of the world. In developing the idea of the "selfish gene," Dawkins was influenced not only by Bill Hamilton but by an evolutionary biologist named George Williams, whose 1966 book *Adaptation and Natural Selection* was a somewhat more technical precursor to *The Selfish Gene*. Williams wrote his book partly in response to the reemergence of the group-selection ideas first put forward by W. C. Allee and Sewall Wright, and in part as a guidebook for determining when sufficient evidence is present to indicate that a trait is the product of natural selection. In his introductory chapter, Williams lays out what he calls his doctrine, namely: "In explaining adaptation, one should assume the adequacy of the simplest form of natural selection, that of alternative alleles in Mendelian populations."[3] In other words, evolutionary biologists—including those studying social behaviors like altruism—should think in terms of genes. Williams's perspective had a profound impact on Dawkins's development of the selfish gene metaphor, and Dawkins makes his gratitude to Williams clear early in *The Selfish Gene*.

Dawkins's gene's-eye view recognizes that natural selection favors any genes that appear to be selfish, in the sense of getting more copies of themselves into the next generation. Most people associate his ideas on natural selection with genes that benefit the individual in which they reside—and no one else. The title of his book might lead one to make such a jump, and it is in fact true that Dawkins's ideas work especially well for genes that benefit only the individual who carries them. But the power of the selfish gene approach is that it works equally well in explaining altruistic traits.

Dawkins realized that Hamilton's kinship theory provided a new way that selfish genes could get copies of themselves into the next generation. With Hamilton's rule in hand, he argued that altruism was the manifestation of the hardworking selfish gene. If a gene can facilitate getting more copies of itself into the next generation by aiding blood kin—individuals that were also likely to possess the very same gene—then altruism easily fits into the selfish gene metaphor. Dawkins was "blown away"[4] by Hamilton's kinship models, and he told his readers that these models "are among the most important contributions to social ethology ever written," while at same time lamenting that when *The Selfish Gene* was published, Hamilton's ideas "have been so neglected by ethologists. . . . His name does not even appear in the index of two major textbooks of ethology."[5]

The notion that altruism was just the long hand of the selfish gene at work would become a working assumption for many in the field of evolution and social behavior, but it was more than Harvard biologist Stephen J. Gould could stomach. For Gould, it was bad enough that people like Williams, Dawkins, and Hamilton were suggesting that many behavioral traits we see are the result of natural selection acting on alternative alleles. He viewed such an approach as "atomistic," relying too heavily on the power of natural selection, and missing the forest (large scale, or macroevolutionary change) for the trees (small scale, or microevolutionary change). But with the publication of *The Self-ish Gene*, this whole gene-centered view of the world was being taken a step further, and altruism between family members—one of the very foundations of morality in humans—was being attributed to these very same selfish genes. Gould and Dawkins would argue on and off about these sorts of issues, until Gould passed away in 2002.[6]

When E. O. Wilson published his now classic book, *Sociobiology*, he was one of the best-known biologists in the United States, having already written a landmark book entitled *Island Biogeography*, which laid out a comprehensive theory for the distribution of species on islands. He was also one of the world's leading entomologists, having summarized his ideas in a monograph called *The Insect Societies* (1971), which is required reading for an anyone interested in insect ecology and evolution.

But it was the publication of *Sociobiology*, a book that made the front page of the *New York Times*, that thrust Wilson into the spotlight. It was essentially a six hundred-page compendium of evidence that social behavior—both animal and human—was a product of natural selection. This idea was not new to ethologists and evolutionary biologists, but it was not familiar to the general public, and it had never been systematically explored in such dramatic fashion before Wilson's book. So, when *New York Times* science correspondent, Boyce Rensberger, published an article entitled "Updating Darwin on Behavior," in which he argued that "sociobiology carries with it the revolutionary implication that much of man's behavior toward his fellows . . . may be as much a product of evolution as is the structure of the hand or the size of the brain,"[7] everyone paid attention—including, once again, Wilson's colleague at Harvard, Stephen Jay Gould, who disliked sociobiology for many of the same reasons that he disliked the selfish gene approach.[8]

Hamilton's rule was to play a significant role in the theories laid out in *Sociobiology*, and there are many folktales about Wilson's initial response to Hamilton's ideas on kinship and altruism. In one version, retold by mathematician and science writer Jim Schwartz, Wilson tossed Hamilton's *Journal of Theoretical Biology* paper in the trash, "believing it to be the ravings of another manic graduate student."[9] Realizing his mistake, Wilson raced from his bed in the middle of the night, frantically going through the garbage in his office before the cleaning lady could whisk it off to the dumpster. Wilson claims that this is nonsense, but his version of events is equally intriguing.

As Wilson tells the story in his autobiography, *Naturalist*, he first read Hamilton's papers during a 1965 train ride from Boston to Miami. He grasped Hamilton's arguments about haplodiploidy, kinship, and altruism in social insects, but found the prose "convoluted" and the mathematics "difficult," and his initial response to the paper was negative. "Impossible, I thought. . . . This can't be right. Too simple. He must not know much about social insects." Wilson, however, believed that Hamilton's papers were worth a second read, during which he searched for the "fatal flaw" that he thought must be embedded in the work. While he looked for the underlying error in Hamilton's model, Wilson tried to develop an alternative hypothesis for altruism in the Hymenoptera. "I modestly thought of myself as the world authority on social insects," he wrote, and so he thought it "unlikely that anyone else could explain their origins, certainly not in one clean stroke." But the more he thought about Hamilton's arguments on kinship and altruism in the Hymenoptera, the more he realized Hamilton was correct. "By the time we reached Miami," Wilson noted, "I gave up. I was a convert, and put myself in Hamilton's hands. I had undergone what historians of science call a paradigm shift."[10]

References to Hamilton's "beguiling" and "audacious" theory of altruism and kinship can be found throughout *Sociobiology*. As one of the most highly regarded entomologists in the world, Wilson was particularly interested in highlighting Hamilton's rule as it applied to social insect behavior—indeed, he praised Hamilton's work in *The Insect Societies*, but that book had a much narrower audience than *Sociobiology*. Recognizing that "the key to hymenopteran success appears to be haplodiploidy," Wilson told the reader that "nothing but kin selection seems to explain the statistical dominance of eusociality [extreme sociality] by the Hymenoptera."[11] In the thirty years since the publication of *Sociobiology*, Wilson's views on the power of haplodiploidy to explain altruism in social insects have changed, but in 1975 he was the lead spokesman for this position.[12]

In chapter 27 of *Sociobiology,* in a section entitled "Role Playing," Wilson tied kin selection theory to one of the most controversial issues raised in his book—the evolution of human homosexuality. "A key question of human biology," he suggested, "is whether there exists a genetic predisposition to enter certain classes and play certain roles. Circumstances can be easily conceived in which such genetic differentiation might occur."[13] In particular, Wilson outlined a hypothesis for the evolution of homosexuality that Herman Spieth had suggested to him and that Robert Trivers[14] had mentioned in a paper the previous year.

Wilson summarized Spieth and Trivers's ideas as follows: "Homosexual members of primitive societies may have functioned as helpers. . . . Freed from the special obligations of parental duties, they could have operated with special efficiency in assisting *close* relatives. Genes favoring homosexuality could then be sustained at a high equilibrium by kin selection alone."[15] In other words, homosexuality could evolve if homosexuals diverted the energy that might normally be used to raise their offspring into helping to raise the offspring of their blood relatives. There is still a raging debate today over whether or not homosexuality has a genetic basis and if homosexuals in fact help raise their blood relatives. Whether in the end the data support or refute Wilson's hypothesis, the critical point here is that in 1975 Wilson believed that Hamilton's ideas on kinship and altruism were so powerful that he used them to speculate on what he knew would be one of *Sociobiology*'s most contentious subjects.

The Selfish Gene and *Sociobiology* instantly became required reading for all evolutionary and behavioral biologists, as well as for the layman science reader. As a result, Hamilton's models of kinship and altruism made their way to a generation of scientists. But it really was these books, rather than Hamilton's original papers that made it all happen. In fact, in a short 1980 article published in the *New Scientist*, Jon Seger and Paul Harvey argued that most biologists did not actually read Hamilton's papers but instead relied on Dawkins's and Wilson's description of inclusive fitness theory. In support of this contention, they found that incorrect citations to Hamilton's *Journal of Theoretical Biology* paper began to appear soon after the publication of Wilson's and Dawkins's books. Over and over they found citations to "The

Genetical Theory of Social Behaviour," rather than to the correct title of Hamilton's papers, "The Genetical Evolution of Social Behaviour." Seger and Harvey discovered that the mutant (incorrect) title could be traced directly back to the bibliography section of *Sociobiology*, in which "Theory" replaced "Evolution." Remarkably, Dawkins had independently made the same error in *The Selfish Gene*.[16]

Much, but certainly not all, of the early work testing the predictions of Hamilton's rule was carried out by those studying what is called "cooperative breeding" in birds.[17] There are many forms of cooperative breeding, but for our purposes, the key aspect of such breeding systems is the presence of what are called "helpers." Such helpers are often young individuals who are physiologically capable of reproduction but choose instead to remain at their natal nest and help raise their siblings. Two of the critical questions that needed to be resolved in such social systems were: why should such altruistic behavior ever be favored by natural selection? and what role does blood kinship play in understanding this phenomenon?

A two-step process has been proposed to explain the evolution of this form of altruism. First, good territories in many bird species are a rare commodity, and, on occasion, the number of territories available to new breeders may become so low that some individuals opt to remain on their birth territory. Second, after an individual has decided to stay at home, it increases its inclusive fitness benefits by helping raise its siblings, in accordance with Hamilton's rule. This hypothesis, however, assumes that having helpers present increases the number of siblings that emerge from nests. But, do nests with helpers actually have greater productivity (that is, produce more chicks) than nests without helpers, and can this be traced to the actions of the helpers?

In principle, it is certainly possible that nests with helpers may fare no better than nests without altruistic helpers. And even if nests with helpers fledge more young, it could be that such nests are simply situated on better territories in the first place. In that case, territory quality per se, rather than the presence of helpers, may explain the increased number of young fledged. The only way to definitively test the effect helpers have on nest productivity is to experimentally manipulate the number of altruistic helpers at a nest, and in so doing experimentally manipulate the variables in Hamilton's rule. In 1976, Jerry and Esther Brown took up that challenge.

During a sabbatical leave from the State University of New York, the Browns traveled to Meandarra, Australia, to construct experiments that would alter the number of helpers on the nests of gray-crowned babbler birds (*Pomatostomus temporalis*). The Browns studied twenty "social units" of babblers. At the start of their experiment, each unit had approximately the same number of individuals—two breeders and four to six helpers. The Browns assigned eleven of these units to a control treatment in which nests were untouched. In the other nine experimental units, the size of the group was reduced to three individuals— the breeding pair and a single helper. The Browns found dramatic differences in the survival of young raised in the control and experimental units. Control units produced an average of 2.4 young, while experimental units averaged 0.8 offspring. Helpers, who because of the paucity of good open territories had little chance of successful reproduction on their own, accounted for an average of 1.6 *additional* offspring—that is, new blood relatives that increased the inclusive fitness of both parents and altruistic helpers.[18] Helpers were altruists and behaved as Hamilton's rule predicted.

In addition to being used to model the evolution of altruism, Hamilton's rule was also soon employed to understand the evolution of sex ratios—that is, to examine what ratio of males to females natural selection might favor in any given population of individuals. To see how, let us return to the social insects that so fascinated Bill Hamilton. The queen of a hive of bees or a colony of ants is always equally related ($r = 0.5$) to her sons and daughters.[19] Long before inclusive theory was created, Fisher had demonstrated that when a female is equally related to her male and female offspring, natural selection should usually favor individuals (in our case queens), who produce 1:1 female/male sex ratio.[20] In social insects, because the queen often lays the eggs for the whole colony, selection should favor queens that set the initial sex ratio of their offspring at 1:1. The catch is that the queen is not the only individual in a hive that can affect sex ratios. By selectively feeding members of one sex over the other, the workers in a hive of social insects can also alter the proportions.

Because workers are three times as related to their sisters ($r = 0.75$) as they are to their brothers ($r = 0.25$), Fisher's model predicts that natural selection should favor workers who manipulate the sex ratio to a three females to one male value. Hamilton's rule, then, predicts a conflict between the queen and the workers over colony sex ratios.

In his 1964 paper, Hamilton had discussed this conflict, but the issue was first brought to the attention of most evolutionary biologists when Robert Trivers and Hope Hare examined social insect sex ratios in a seminal 1976 *Science* paper entitled "Haplodiploidy and the Evolution of the Social Insects."[21] Trivers and Hare used published data on twenty-one species of ants to examine whether the sex ratio battle was being won by workers or by queens. What they found was a clear victory for the working class: the sex ratio in all the species that they studied was not statistically different from three females to one male. And while on the surface this seems to be all about sex ratio evolution, the Trivers and Hare study touches on kin selection and altruism as well, for by controlling the sex ratio in a colony, worker ants are fashioning colonies that are stock full of their altruistic sisters ($r = 0.75$).[22]

Jerry and Esther Brown, Robert Trivers, and Hope Hare were only a few of the people who would put Hamilton's ideas to the test—some came before them and hundreds more would follow. Looking back on his work on kinship and altruism, Hamilton once mused, "I like always to imagine that I and we are above all that, subject to far more mysterious laws. In this prejudice, however, I seem, rather sadly, to have been losing more ground than I gain. The theory I outline . . . has turned out very successful. It certainly illuminates not only animal behavior but, to some extent as yet unknown but actively being researched, human behavior as well."[23] In order to see just how illuminating Hamilton's ideas on kinship and altruism have become, we now turn our attention to a group of researchers at Cornell University.

Keepers of the Flame

I F THERE IS a focal point for modern work on kinship and altruism, it resides on the third floor of Seeley Mudd Hall at Cornell University. This place houses a group of scientists that, as a unit, have done more work testing Hamilton's rule than any group, anywhere. Though they shy away from thinking of themselves as having a leader, that distinction falls on the shoulders of the most senior of these scientists, Stephen Emlen. In addition to seniority, Emlen has the added qualification of being part of one of the few true dynasties in the area of behavior and evolution: his father, John T. Emlen, is considered a founder of animal behavior and modern population biology; his brother, John Merritt Emlen, is a recognized population and community ecologist; and his son, Douglas Emlen, has made a name for himself studying the development and genetics of behavior at the University of Montana.

Stephen Emlen first read Bill Hamilton's papers on kinship and social behavior in the mid-1960s when he was one of Richard Alexander's graduate students at the University of Michigan.[1] Emlen's dissertation research was on migration behavior in birds, but Hamilton's kin selection models struck him as immensely important—an "aha" moment, as Emlen recalls it. And so while Hamilton's rule did not make a big splash until the mid-1970s, it seriously influenced Emlen a decade earlier. Busy with other projects, though, he tucked Hamilton's ideas away in the back of his mind until he had time to do something more with them. Soon after completing his dissertation, he accepted a job as an assistant professor in the department of neurobiology and behavior at Cornell. Emlen returned to Hamilton's models of kinship and altruism when he used his first sabbatical leave from Cornell in 1973 to develop a research program focusing on the evolution of cooperative breeding in birds.[2] Emlen's Nigerian colleague, Hiliary Fry, suggested to him that he work with white-fronted bee-eaters, a cooperatively breeding species native to Kenya.[3] This little bird would turn out to be a treasure chest of information on kinship and altruism—one that Emlen would tap into many times over the years.

White-fronted bee-eaters nest on the face of vertical, dirt-covered cliffs. Hundreds of individuals each dig meter-long tunnels, about fifteen to twenty centimeters apart, and build their nests at the end of these tunnels.[4] The altruists in this species are young "helpers" who assist breeders in digging and defending nests, and who also feed the breeders and any chicks they raise—and, the beauty of the system is that almost all of this can be recorded by the naked eye. This combination of factors could not be found in any of the local species near Cornell, and it made the bee-eaters irresistible. So, despite the fact that he had absolutely no experience doing field work outside of the United States, in the spring of 1973 Emlen, along with his wife, Natalie Demong, headed to Lake Nakuru National Park in Kenya to study these birds, in part to see whether Hamilton's rule could help explain their extraordinarily cooperative social system.[5]

Studying kinship and altruism in bee-eaters is a labor-intensive job. Each year in the early spring, Emlen and his team of assistants arrive and mark all the birds that have not already been tagged with a metal leg and wing bands. Once marking is complete, Emlen and his team spend their days sitting behind a small "blind," looking out at bee-eater colonies with hundreds of nest holes. They refer to each hole as an "apartment," and each apartment has an "address" (for example, hole 22a). Fortunately, because the nests are extremely narrow, almost all behaviors of interest—who brings in food, what sort of food they bring in, who fights with whom, who wins a fight—take place at the nest entrance, where they can be recorded. Each day Emlen and his assistants choose a different subset of apartments to watch and record everything they observe into handheld tape recorders. After nightfall, they enter all of the day's observations into a computer. "You have these dossiers on each bird," Emlen says, "like the CIA."[6] These dossiers not only enable Emlen to know who is related to whom, but allow for the behavior of a specific bee-eater, altruists included, to be compared to what that same individual did in the past, or what other birds in the colony are doing.

Emlen has tested a number of predictions associated with

Hamilton's theory on kinship and altruism. For example, Hamilton's rule clearly predicts that individuals should prefer to help blood kin to whom they are closely related over more distant blood relatives. To test this prediction, Emlen used 115 instances he recorded in which a bee-eater had the option of helping two individuals that differed in their degree of blood relatedness; the coefficient of relatedness (r) between pairs ranged from 0 (unrelated) to 0.5 (full siblings or parent/offspring). Helpers chose to aid their closest kin on a remarkable 94 percent of these occasions.[7]

In addition to testing a basic tenet of kinship theory, Emlen's results helped resolve a thorny issue surrounding Hamilton's rule. Starting in 1975, a number of researchers, including Mary Jane West-Eberhard at the Smithsonian Tropical Research Institute and David Barash at the University of Washington, had suggested that individuals should dispense altruistic aid to relatives in *direct proportion* to blood relatedness (the "proportional altruism" model). For example, imagine an individual with nine units worth of aid that it can dispense to relatives. Suppose that this individual consistently interacts with one sibling ($r = 0.5$) and one uncle ($r = 0.25$). Since siblings share an r value twice as great as that between uncle and nephew, the proportional altruism model predicts that six units of aid should be dispensed to the sibling and three units of aid should be dispensed to the uncle.

Not everyone thought that the proportional altruism model was correct. For example, the University of Chicago's Stuart Altmann claimed it was based on faulty logic, because an individual always increases its inclusive fitness most when it is altruistic toward its closest blood relative. As such, Altmann argued that an individual should dispense *all* of its aid to the recipient that is its closest blood relative (the "all-or-nothing" model). In our hypothetical case, Altmann's model predicts that all nine units should be dispensed toward the donor's sibling. In principle, Altmann is right, but the question really was whether animals behave in accordance with Altmann's predictions, and that is where Emlen's work on white-fronted bee-eaters comes in, for it allowed behavioral and evolutionary biologists to determine which of these two models better fits data gathered in the wild. In support of Altmann's model, Emlen found that helpers not only overwhelmingly chose to help their closest blood relative, but that once a helper made a choice, it also dispensed all of its aid toward the chosen individual.

Emlen's studies on cooperation expand the reach of Hamilton's rule and go beyond experiments on cooperative breeding and kinship in white-fronted bee-eaters. In the 1980s and 1990s, he developed a series of models that examine how ecology and blood kinship jointly influence altruism and other forms of social behavior, and these models were summarized in a 1995 review paper entitled "An Evolutionary Theory of the Family." Emlen's theory of the family includes variables representing genetic relatedness, (r), dispersal options available to mature offspring, and the competitive abilities of all individuals involved. These different components all tie back rather nicely to Hamilton's rule ($r \times b > c$), for, in essence, the dispersal options and competitive ability correspond to the costs and benefits that Hamilton captured with his b and c variables. That said, translating Hamilton's b and c into measurable, evolutionarily relevant behaviors was no small task, and Emlen's model has generated some new, interesting questions for those interested in kinship and altruism.

In his 1995 paper, Emlen made fifteen specific predictions about "the formation, the stability and the social dynamics of biological families."[8] For each of these predictions, he reviewed the *animal behavior* literature in search of evidence both for and against them. Then, two years after Emlen's review article was published, Jennifer Davis and Martin Daly of McMaster University used data from the 1990 Canadian General Social Survey (CGSS)[9] to test Emlen's fifteen predictions as they relate to human family dynamics.[10] Together, these two studies provide a unique opportunity to study kinship, altruism, and an evolutionary theory of family in both humans and nonhumans. To see how, let us explore three of Emlen's predictions in more depth.

The first relates to the stability of the family unit, namely, that "family dynamics will be unstable, disintegrating when acceptable reproductive opportunities materialize elsewhere."[11] The logic here is straightforward; Hamilton's rule predicts that helping genetic relatives is one means to pass on genes, but there are often other, more direct, ways to do that—like having your own offspring. If those more direct ways provide a better means for getting genes into the next generation, they should be employed, even if it leads to the disintegration of the existing family unit and a concomitant decrease in familial altruism in that family.

To test this hypothesis, Emlen relied on studies in which researchers experimentally created new, unoccupied territories and then examined whether mature offspring would emigrate from their parents' home to such newly formed territories. Results from nine long-term studies involving acorn woodpeckers, red-cockaded woodpeckers, green woodhoopoes, superb fairy wrens, Florida scrub jays, Galapagos mockingbirds, Seychelles warblers, white-footed mice, and kangaroo rats support his prediction. As an example, in superb fairy wrens (*Malurus cyeneus*), young males often help their parents. To test the idea that the family unit—and hence altruism among kin—breaks down when suitable territories for such young helper males emerge, Stephen Pruett-Jones from the University of Chicago created new breeding opportunities by removing breeding males from twenty-nine superb wren territories.[12] He found that thirty-one of the thirty-two male helpers who could have dispersed from their parent's nests to the newly opened territories did so, and they did so quickly—new territories usually were occupied by former helpers within six hours.

When testing Emlen's first prediction regarding family stability in humans, Davis and Daly found that married individuals were much *more* likely to live away from their parents than were single individuals in the same age and sex category, suggesting that new marriages cause existing family units to dissolve. It need not have turned out that way; married individuals could just have easily lived with one set of in-laws. It is certainly the case that cultural norms in Western society encourage young married couples to move away from home and establish independence, but the real question then becomes why such cultural norms exist in the first place, and that is where a biological theory of blood kinship and altruism becomes very useful.

The fact that married individuals were more likely to live away from their parents than were single individuals is a very coarse measurement of "family stability," as it looks at physical location as the sole measure of cohesiveness. In our own species, contact between parent and offspring need not end when children leave home. Family members can stay in contact by phone calls, letters, visits, or e-mail, and so Davis and Daly examined whether married children who left home were more or less likely than unmarried children who left home to keep in touch with their parents. They reasoned that if families truly dissolve when offspring leave home to start their own families, then married children living away from home should have *less* contact with their parents than single children living away from home. What they found, however was that for most age and sex categories, married individuals living apart from their parents were just as likely to stay in contact with parents (and grandparents) as single individuals living away from home.[13] Based on the results of their analyses, Davis and Daly argue that the data from the CGSS do not support Emlen's prognosis: "At best Emlen's first prediction," they note, "holds only for some demographic groups with some types of relatives."[14]

They hypothesize that part of the reason that the data from humans do not provide strong support for Emlen's hypothesis centers on the fact that we need to rethink the idea of "helpers" when we deal with humans. One of the Emlen's assumptions is that young adults are the ones acting as helpers. That is, while they live with their parents, young adults act as helpers, but once they marry, their effort is shifted from aiding their parents to raising their own family. It turns, out, however, that while this is certainly true for humans, it is also the case that, unlike most other animals, once young adults marry and move out, their parents can act as helpers and provide much-needed resources (money, time), which all leads very nicely to Emlen's next prediction.

Instead of asking what happens when breeding opportunities away from home become available, Emlen's second prediction addresses the question of whether changes in the level of resources available on a territory affect the stability of families, and hence the amount of familial altruism. More specifically, he suggests the following: "Families that control high quality resources will be more stable than those with lower quality resources. Some resource-rich areas will support dynasties in which one genetic lineage continuously occupies the same area over many successive generations."[15] In the language of Hamilton's rule, if the inclusive

fitness benefits of staying at home are great enough, natural selection should favor very strong family bonds—strong enough
that offspring remain on their birth territory for their entire
lives.[16] Offspring help, mature, choose mates, and reproduce all
in the same area.

Data on family stability in acorn woodpeckers, Mexican jays,
Florida scrub jays, striped-backed wrens, and Seychelles warblers all support Emlen's "dynasty building" prediction. In
these five species of birds, individuals raised on high-quality natal territories were much less likely to leave home than individuals raised on inferior territories. For example, in cooperatively
breeding acorn woodpeckers (*Melanerpes formicivorous*), the critical measure of a territory's quality is the number of storage
holes available for acorns. Acorn woodpecker territories vary
from having fewer than one thousand to greater than three thousand storage holes, and Peter Stacey and John Ligon from the
University of New Mexico found that individuals on territories
with lots of storage holes produce a greater average number
of offspring. More important for Emlen's second prediction, on
territories with more than three thousand storage holes, 27 percent of the young stayed and provided altruistic aid to their
relatives—this despite the fact that many open territories were
available.[17] The corresponding figure for territories with fewer
than one thousand holes is less than 2 percent.[18]

In terms of resource levels and family stability in our own
species, Davis and Daly found that young adults from rich families tend to be *less* likely to be living with their parents than same-
aged individuals from poorer families.[19] This is clearly the opposite of what one would expect from Emlen's prediction. However,
unlike the number of acorn holes on a territory, economic resources in humans are portable, so familial coresidence may be an
inappropriate yardstick for measuring family stability. If family
stability is defined instead in terms of maintaining family contacts
during adulthood, Davis and Daly found that such contact is indeed more often found in wealthy families than in poor families.[20]

Emlen's third prediction is another extension of Hamilton's
rule, which asserts not only that altruism should vary with genetic relatedness, but that when $r = 0$ (individuals are not blood kin),
altruism should be rare. As such, Emlen claims that "replacement
mates (stepparents) will invest less in existing offspring than will

129

biological parents."[21] That is, when stepparents are faced with decisions about how to distribute resources to offspring, they should favor their own biological offspring ($r = 0.5$) over stepchildren ($r = 0$).[22] Data from rodents, carnivores, and primates all support Emlen's prognosis that critical resources such as food and protection should be dispensed to biological offspring over stepchildren.[23] Not only are resources withheld from stepchildren more often, but infanticide on the part of male, as well as female, stepparents has also been uncovered in a number of species.[24]

Cross-cultural data on humans clearly demonstrates that stepparents invest less in their stepchildren than in their biological children.[25] More disturbingly, comparisons of biological versus stepfamilies also find that stepchildren not only receive fewer resources, but also suffer child abuse at rates up to one hundred times higher than those of biological children.[26] In their book, *The Truth about Cinderella: A Darwinian View of Parental Love*, Daly (of Davis and Daly) and Margo Wilson (Daly's colleague and wife) described the situation in stark clinical terms: "Having a step-parent has turned out to be the most powerful epidemiological risk factor for severe child maltreatment yet discovered."[27] This maltreatment includes the most severe forms of child abuse, in that most large-scale studies demonstrate that stepchildren are murdered at a much higher rate than biological offspring.[28] Such abuse and murder are predominately at the hands of the stepparent (versus the biological parent), and because custom and law around the world favor a mother's rights, the abusing parent is almost always a stepfather.[29]

If we take Hamilton's rule one step further and employ it when we study *adopted* children and child abuse, a clear prediction emerges. Adopted children have two stepparents, which simply means that the blood relatedness between adopted children and *both* their new parents is zero, so altruism should be even less common here than in the stepparent case, and, conversely, child abuse should increase. The available evidence from the United States and Canada, however, suggests just the opposite: adopted children suffer much less child abuse than any other group of children. But there is a good reason for that, and this is where cultural forces come into play.

In the United States and Canada, all prospective parents must go through a stringent selection process in order to adopt. This involves long interviews with experts working for the adoption agency, as well as the screening of applicants with respect to family history, history of violence, and financial stability—which is to say that these countries have set up cultural norms specifically designed to protect adopted children against violence. Such norms serve as a form of protection against our inherent tendencies to treat nonblood kin—in this case, adopted children—differently from blood kin. The sad truth, though, is that in many other societies, particularly non-Western societies, these cultural norms simply do not exist, and the screening process for adoption is lax, if present at all. Ethnographic data show that when these protections are absent, child abuse toward adopted children increases dramatically. Indeed, Margo Wilson notes, "if you look historically and cross-culturally wherever there are few government mechanisms to protect children then there are lots of horrific stories of abuse and exploitation of unrelated adoptees."[30] In other words, remove the *cultural norms* associated with adoption in places like the United States and Canada, and violence directed at adopted children looks eerily like the case of child abuse directed at stepchildren. What seems clear then is that cultural rules can indeed curb the inherent biases toward kin that Hamilton's rule predicts, but we often fail to implement such rules.

The societal implications of both Hamilton's rule, and the extensions of it, have not escaped Emlen. Indeed, he argues that his evolutionary models of stepfamily violence, for example, are "empowering" and that he has discussed these models with many social scientists. Such discussions can be tricky, because social scientists think about behavior in a very different way than do evolutionary biologists. Evolutionary biologists are trained to ask how natural selection favors one behavior over another. Social scientists, on the other hand, tend to ask how developmental issues (family history, upbringing, and so forth) affect behavior. Therapists are interested in "what happened to you when you were growing up," Emlen notes, "talking things out rather than coming out front and saying 'statistically, you are in a high-risk group.'" In addition to discussing developmental issues, he believes that therapists need to tell their patients the following: "The logic of the evolutionary approach is . . . [that] you shouldn't feel guilty that you have a hard time bonding with your stepson. That is probably totally normal. In order to have a good functioning family you have to work all the harder in your actions with your stepson as compared to your real son." Emlen has tried out this approach at a number of major social science conferences. Overall, the reaction of social scientists has been positive, but as he soberly notes, "those who thought it sucked may have walked out the back door."[31]

One of Emlen's long-time colleagues in the department of neurobiology and behavior at Cornell University is Paul Sherman—no stranger himself to working with kinship and altruism. Indeed, Sherman is often credited with doing the first direct empirical test of Hamilton's rule in the mid-1970s, when in the September 25, 1977, issue of *Science* he reported the results of his dissertation work[32] on alarm calls and blood kinship effects in Belding's ground squirrels (*Spermophilus beldingi*).[33]

As part of his graduate work, Sherman had spent more than three thousand hours watching ground squirrel behavior at the Tioga Pass Meadow in the Sierra Nevada. Over the course of his observations, he noticed that when a ground squirrel spotted a predator, it would often stand up on its hind legs and emit a piercing scream. Every squirrel in the vicinity, including the caller, then ran for cover. Sherman recognized that not all squirrels were equally likely to emit such alarm calls. But why? Why were some squirrels more likely than others to sound these altruistic warnings—calls that drew attention to the themselves but benefited others in their group? To answer that question requires knowledge about the demography of Belding's ground squirrels.

The males of the species move to new populations when they mature, but the female squirrels spend their entire lives in their population of birth. This difference in dispersal creates an asymmetry in the way that adult males and females are related to others living in their groups. By remaining in the populations in which they were born, females—both young and old—are always surrounded by blood relatives. Mature males, who emigrate to new populations, however, find themselves interacting with complete strangers. Movement patterns affect whether these squirrels interact with blood kin when they are adults, and this then sets the stage for introducing Hamilton's rule into the mystery surrounding who gives altruistic alarm calls.

"Ground squirrels are like tulips," Sherman muses. "They come up in the spring, they bloom, and then they go down. . . . You can watch them, catch them, mark them, and they do most of what they do on the surface during the day."[34] Each spring, Sherman and an army of field assistants began their field season by capturing and marking every untagged squirrel in the Tioga Pass population. After all the marking was complete, he and his team spent their days gathering data on every ground squirrel behavior imaginable. Pages of Sherman's field notebooks, as well as those of his assistants, are filled with data on chases, fights, territory size, mating, and so on. Their routine was interrupted by only two things—the appearance of an unmarked squirrel (which Sherman's team then raced to capture and mark) and the appearance of a predator. Altruistic alarm calls come in response to predators, so the research team kicked into high gear when a chance to record data on such calls emerged.

Once a predator appeared, Sherman and his team scurried about trying to take notes on which squirrels were present and which were not, and who was calling, running, hiding, and so forth. This sort of data was next to impossible to obtain for aerial predators, who strike with lighting speed, but was gathered fairly routinely for terrestrial hunters. Indeed, over the course of his three-year study, Sherman saw terrestrial predators and squirrels together on 102 occasions, and the weasels, badgers, dogs, coyotes, and pine martins that he observed killed a total of six adult Belding's ground squirrels and three juveniles.

Sherman discovered that when a terrestrial predator was spotted, female squirrels gave alarm calls much more often than expected by chance, but the converse was true for males. On average, when a predator was spotted by ground squirrels, 30 percent of the individuals present were adult females and 20 percent were adult males; the rest were juveniles and "subadults." But when Sherman analyzed which squirrel sounded the first call, he found that females did so about 65 percent of the time (more than twice the 30 percent that would be expected by chance alone), but males did so 2 to 3 percent of the time (about one-fifth to one-tenth of what would be expected by chance). Adult females, surrounded by blood kin, were emitting the altruistic alarm calls, while adult males, unrelated to those around them, kept quiet. Sherman, who had learned about Hamilton's rule in graduate school, realized that alarm callers were aiding their blood relatives, just as Hamilton's rule predicted they should.[35] Further support for the hypothesis that blood kinship helped explain who gave alarm calls emerged when Sherman discovered that in the rare instance when a female did leave her natal group and moved into a population of unrelated individuals, she gave alarm calls less frequently than did native females.

Aerial predators attack ground squirrels so quickly that over the course of eight field seasons, Sherman had observed only an average of seven attacks per year. He knew he would need a large sample size to test whether Hamilton's rule helped explain alarm calls in response to aerial predators, and so he decided to manipulate the frequency and timing of aerial attacks himself. He hired two falconers, Gary Beeman, who owned a falcon named Chapparal, and Jeff King, who owned a bird named Mojo. He then placed his assistants at various positions around a field, each to watch one squirrel. As each observer locked on to his or her respective squirrel, he or she raised one hand in the air. When everyone's hand was up, Sherman used a walkie-talkie to order the falconers to release either Mojo or Chapparal. "The hawk would lift and do a straight line between the guy who was holding the hawk and the guy who had a chicken leg the hawk wanted," Sherman notes, "and then data would come. And we'd do it again, and again and again."[36]

Sherman's data on "whistle" alarm calls directed at aerial predators is in stark contrast to that for alarm calls associated with terrestrial predators. When danger descended from the air, the distribution of alarm calls was *not* associated with blood relatedness, and, in particular "females' tendencies to whistle were not affected by the presence of relatives."[37] The reason the presence of blood kin did not affect alarm calls in response to aerial predators turns out to be very simple. When a call is given in response to a terrestrial predator, the caller puts itself in danger—the call is truly altruistic—and hence Hamilton's rule suggests that there must be some compensating benefit. That benefit is the safety conferred on blood kin. But alarm calls in response to aerial predators actually increase the survival probability of the caller, because aerial hunters, once sighted, have little chance of success, and veer off toward other prey. In such a case, there is no need to search for indirect benefits that blood relatives might receive to explain their evolution—standard selfish gene models do the trick here.

Sherman's work on inclusive fitness garnered tremendous attention, and many job offers followed. As he recalls it, "people were enthusiastic and excited about my work.... They knew I was a skeptical scientist.... I wasn't blindsided by 'oh, here is a new concept and we will have to fit everything into it.' "[38] What Sherman did show, though, was how Hamilton's rule could be experimentally tested in a complex, yet manipulable system. For some behaviors—terrestrial alarm calls—it explained what was otherwise difficult to explain in nature. For others—alarm calls directed at aerial predators—Hamilton's models were not necessary to explain what was happening in the wild.

In retrospect, an interesting question arises from the Sherman study: Although the results of his work were clearly in line with the predictions of Hamilton's models, and Sherman did know the genetic relatedness (r) of the squirrels in his population, he never had detailed measurements of the b and c terms in Hamilton's rule—instead he had qualitative measures of the benefits and costs of alarm calls. As such, one might ask whether Hamilton's equations were necessary for the sort of experiments that he undertook. Would not Haldane's and Fisher's verbal models have sufficed to stimulate such work? The answer to this question appears to be no. In the first place, such verbal models had been around for forty years and had not spurred on such work. More to the point, Sherman and future researchers are all quite adamant that it was Bill Hamilton's ideas that were the impetus for their studies. Devising a simple equation with three variables that captures kinship and altruism made all the difference, and as a result of Sherman's work on ground squirrels, Hamilton's rule on altruism and kinship went from an abstract theory to a testable model.

Thirty years after his seminal work on altruism and blood kinship, Sherman's passion for studying this subject has not diminished. He has, however, switched species and focused on the implications of Hamilton's rule for the buck-toothed, hairless, virtually blind naked mole-rat (*Heterocephalus glaber*). In evolutionary circles, the story of naked mole-rats and Hamilton's rule is legendary and begins at the University of Nairobi with a scientist named Jenny Jarvis.

As part of her dissertation work on the mole-rats of eastern Africa, Jarvis began working with the naked mole-rat in 1967. Shortly later, she realized something very odd, which was that she rarely encountered a pregnant naked mole-rat. Not sure what to make of her finding, Jarvis continued her field work and also secured a faculty position at the University of Cape Town, where she attempted to form naked mole-rat lab colonies by mixing individuals caught in different wild colonies. At first, the females were very aggressive toward one another. Eventually, however, the fighting stopped, and a single female, often referred to as the queen, became the sole breeder in a colony. A single queen per colony helped explain why Jarvis had problems finding pregnant females, but was nonetheless totally unexpected; bees and other social insects, not mammals, are supposed to have queens.

In the spring of 1976, unaware of Jarvis's (unpublished) work on naked mole-rats, Richard Alexander was traveling across the western United States, lecturing on kinship, altruism, and evolution. Alexander was particularly fascinated with an extreme form of altruism called "eusociality." A eusocial society is defined as one in which there is a reproductive division of labor (queen and workers), overlapping generations (parents and offspring have lifespans that significantly overlap), and communal care of young. One of Alexander's lectures was entitled "Why Are There No Eusocial Mammals?" Why, he, asked, was eusociality apparently limited to insects? Haplodiploidy in the social insects, and the extremely high genetic relatedness among workers and the altruism that resulted from it, was clearly part of the explanation. But haplodiploidy could not be the whole answer, Alexander argued, because termites are eusocial, and they are diploids (like mammals), not haplodiploids.[39]

What makes termites special is that even though they are not haplodiploid, the average genetic relatedness in termite colonies is very high—not as high as the $r = 0.75$ often found among haplodiploid sisters, but high nonetheless. Using the termite as his archetype, Alexander sketched out a hypothetical eusocial mammal. Such a mammal, he speculated, would be part of a colony full of blood relatives, live in very safe, expandable underground nests, and feed on large, abundant tubers that could be stored for long periods of times. This hypothetical creature, then, would lead a mammalian version of termite life. In evolutionary terms, Alexander was hypothesizing that blood kinship and just the right ecological conditions could produce a eusocial mammal.

Alexander presented his "eusocial mammal" lecture at Northern Arizona University in May 1976, and sitting in the audience at that seminar was a scientist named Terry Vaughn. When Alexander was finished with his talk, Vaughn told him, "Your hypothetical eusocial mammal is a perfect description of the naked mole-rat of Africa."[40] Vaughn then showed Alexander some preserved naked mole-rat specimens and provided him with contact information for Jenny Jarvis. The stage seemed set for finding a superaltruistic mammal, surrounded by kin, including the queen of the colony.

At about the same time that Alexander was in Arizona learning about naked mole-rats, his former Ph.D. student, Paul Sherman was attending the annual meeting of the Society of Mammology. There he met a physiologist-anatomist named Milton Hildebrand, who had given a talk about a strange creature called the naked mole-rat. For added effect, Hildebrand even showed his audience, including Sherman, a freeze-dried naked mole-rat. "Dick [Alexander] and I would get on the phone with each other and I'd say "guess what I heard last week," Sherman recalls, "and then he'd say 'Ah, shut up, I gotta tell you what I heard last week.' And we said, 'oh my god, we heard the same thing.'" Alexander and Sherman became more and more fascinated with naked mole-rats, and in November 1979, they packed up their bags and went to Cape Town to visit Jenny Jarvis. Sherman remembers that Jarvis, a physiologist by training, was a bit leery "that these Americans were going to come over and move in on her."[41] After just a few days, however, her fears disappeared, and she welcomed the potential collaboration with colleagues who took a more evolutionary view of naked mole-rat life.

While working with Jarvis, Alexander and Sherman captured naked mole-rats and brought them back to start colonies at the

University of Michigan and Cornell University. A few years later, one of Sherman's graduate students, Kern Reeve, worked with the Cornell naked mole-rat population for his doctoral work on kinship, cooperation, and aggression. Reeve's experiments were so impressive that by 1994 he joined his old mentor, Sherman, on the faculty at Cornell.

First collected in Ethiopia in 1842, naked mole-rats live in underground burrows that house about seventy individuals. These burrows average about two miles in length and are often found in arid areas covered with dust and brush. Typically located near dirt roads, colonies are visible from above by "molehills" that pock the landscape. The queen has a twofold advantage over other females in a colony. She not only monopolizes all colony reproduction, but she also lives longer than her nonre-producing colony mates.[42] Most of the everyday chores associated with colony life are handled by the nonreproductive male and female "workers." Sherman, Reeve, and their colleagues have found that such workers are extraordinarily altruistic; they dig new tunnels for the colony, sweep debris away, groom one another and the queen, gather food for the colony, and defend the colony from predators.[43]

In a 1981 paper published in *Science*, Jenny Jarvis convincingly argued that naked mole-rats were indeed eusocial mammals.[44] Eusociality in naked mole-rats has evolved as the result of two factors: ecological constraints and blood kinship. Jarvis hypothesizes that early on in mole-rat evolution, severe droughts selected for group-living and cooperation. In the language of Hamilton's rule, these droughts (and their effects) help us understand the b and c terms associated with altruism. Droughts may have been so severe that only mole-rat populations in which the young remained in their natal colony, and helped their relatives, survived. Once environmental conditions selected for a group-living species in which individuals remained at home to aid their blood relatives, helpers could start accruing inclusive fitness benefits. But the extent of such benefits—in part measured by the r term in Hamilton's rule—remained unknown until 1990, when Reeve, Sherman, and their colleagues measured genetic relatedness in naked mole-rat colonies.[45]

Using DNA fingerprinting, Reeve and his team discovered that the average genetic relatedness among individuals in naked mole-rat colonies in nature was astonishingly high—0.81 (a value close to the $r = 1$ of clones, and higher than the 0.75 that exists between social insect sisters). This r value—the highest ever recorded for a natural population of mammals—is the result of significant amounts of inbreeding within naked mole-rat colonies. That is, while hymenopteran sisters rely on the strange genetics of haplodiploidy to create high r values, naked mole-rats manage to achieve an even higher r score through blood kin mating with each other. Indeed, Reeve and Sherman estimate that more than 80 percent of all matings in a naked mole-rat colony are between siblings or parents and offspring.[46]

The high r value found in the species, paired with the high benefits associated with naked mole-rat altruism, have been touted as a classic example of Hamilton's rule in action. High r values, however, do more than explain naked mole-rat cooperation; they also shed light on naked mole-rat aggression. Reeve has found that queens tend to "shove" lazy workers, and that such shoves increase the productivity of these workers.[47] But lazy workers are not a random subset of individuals in a colony; they tend to be those workers who are least related to the queen.[48] As predicted by Hamilton's rule, when the effect of kinship is weakened, the degree of altruism decreases, and conflict results.

In addition to working together on the evolution of naked mole-rat kinship and altruism, Sherman and Reeve share a second common interest relevant to Hamilton's rule—understanding how individuals recognize one another as kin. The better an individual is at distinguishing kin from nonkin, the more likely Hamilton's rule is to predict altruism, for the benefits of altruism can then be more precisely directed at blood kin.[49]

There are a number of different mathematical models of "kin recognition" in the literature on evolution and behavior. One such model assumes that every individual generates what is called an "internal template."[50] Such a template might be generated by an individual smelling the nest it lives in and then creating an olfactory template based on that smell. Once individual 1 generates an internal template, he or she then determines if individual 2 is kin or nonkin depending on how closely individual 2 matches that template.[51] For example, in the case of a template generated by odors, individual 2 might be considered kin or nonkin depending on how closely his own odor matches the odor underlying individual 1's internal template.[52]

In collaboration with Reeve and Sherman, David Pfennig has uncovered an interesting case of template matching in spadefoot toad tadpoles (*Scaphiopus bombifrons*).[53] In nature, two different spadefoot toad tadpole "feeding morphs" exist: some juveniles feed on detritus (small, often drifting vegetative clumps of food) and develop into herbivores as adults, while others feed on shrimp and typically mature into carnivorous, cannibalistic adults.

Pfennig, Reeve, and Sherman examined kin recognition abilities in the cannibal morph of the spadefoot toad. Given a choice between ingesting siblings or unrelated individuals, they found that carnivorous morphs were more likely to eat unrelated individuals. Furthermore, the foraging behavior of the carnivorous morphs strongly suggested that they gauged genetic relatedness by using a template-matching rule based on taste: carnivorous morphs were just as likely to suck relatives and nonrelatives into their mouths, but relatives were spit back out much more often. As a result of strong selection pressures associated with maximizing inclusive fitness, spadefoots can actually "taste" genetic relatedness.

Template matching is not the only way that blood relatives can recognize one another. If individuals live in kin groups that are spatially distinct from one another (for example, nests in a forest), a second form of kin recognition can also evolve. When spatial segregation between family groups exists, Sherman and his colleagues have argued that natural selection might favor a kin recognition rule such as "if it lives in your nest/cave/territory, then treat it like kin."[54] Sherman and John Hoogland studied such a kin recognition rule in the bank swallow (*Riparia riparia*). Bank swallow chicks cannot fly until they are about three weeks old, and during this period, parents adopt the following rule: "if it is in my nest, it is likely kin." As a result, they feed any chicks in their nest. After bank swallow chicks learn to fly, parents no longer employ this strategy but instead rely on the distinctive vocal cues produced by each chick to recognize their offspring. This switch in kin recognition makes perfect sense in light of Hamilton's rule.[55] Before chicks are three weeks old, parents can be certain that any chick in their nest is their blood kin, and so all the food brought back to a nest by a parent is dispensed to genetic relatives. Once chicks can fly, however, parents can no longer be certain that the young in their nest are their own, and so they must switch to a more fine-tuned kin recognition rule.

Together, Emlen, Sherman, and Reeve have performed dozens of experiments inspired by Hamilton's rule. More than that, they have expanded Hamilton's original ideas by emphasizing the manner in which ecology and kinship interact to shape behavior. As a result of their work, evolutionary and behavioral biologists now view Cornell University as the home of modern kinship theory. And while Cornell was establishing its program in altruism and kinship, Bill Hamilton too was busy—both developing new models and spending time with his beloved insects.

Curator of Mathematical Models

WHEN E. O. WILSON'S *Sociobiology* was published in 1975, Bill Hamilton was once again studying insects in the forests of Brazil, this time with his wife, Christine, and their two daughters by his side. When Hamilton returned to England in 1976, he realized that twelve years after starting at Imperial College, and despite the fact that his papers on altruism and kinship were now being recognized as seminal, he still only held the relatively low rank of lecturer. The reasons for his nonpromotion are unclear, but it stemmed at least in part from the fact that although the quality of his publications was superb, his rate of publishing was relatively low. And he did not help himself by being off in Brazil, rather than socializing with those in his department. Frustrated by the lack of recognition at home, Hamilton wrote a series of letters to his colleagues in the United States to determine if anyone knew of an open position that might suit him. An enthusiastic reply came from the University of Michigan's Richard Alexander, who was touring the country giving his "eusocial mammal" lecture. Alexander and his colleague Don Tinkle created a position for Hamilton in both the biology department and at the Museum of Zoology.

After a six-month stopover as a visiting professor at Harvard University, Hamilton began his new job at the University of Michigan. Technically speaking, his position at the Museum of Zoology was a five-year editorship of the museum's publications, which Hamilton described as "managing commas, italics and silverpoint plate proofs." Indeed, as he traveled the country giving lectures, Hamilton would tell "puzzled listeners" in the audience that he "curated the museum's mathematical models."[1] On paper, Hamilton's teaching duties at Michigan were fairly light, and he was required to teach only graduate classes. He relished teaching what was widely regarded as the best group of graduate students in the country, and he did his best and most innovative instructing there.

In 1977, Hamilton met a University of Michigan political

scientist named Robert Axelrod.[2] At the time, Axelrod was developing a game theoretical computer simulation of altruism and cooperation between unrelated individuals—work that would eventually earn him a MacArthur Genius Award and election into the National Academy of Sciences. In particular, he was using a game called the prisoner's dilemma to study cooperation and altruism.[3] Axelrod knew of Hamilton's work on inclusive fitness, kinship, and altruism from having recently read Dawkins's 1976 book *The Selfish Gene*, but he had no idea that Hamilton was also at the University of Michigan.[4] Fortunately, he had recently spoken with Dawkins, who informed him of Hamilton's whereabouts. Axelrod then phoned Hamilton and told him that "the situations that he was simulating by computer were getting to be a bit biological" and suggested that the two of them meet and exchange ideas on the evolution of cooperation and altruism.[5]

Although he had done some game theory modeling,[6] Hamilton was keen to learn more about the prisoner's dilemma and cooperation between unrelated individuals, and he was delighted at the opportunity to meet Axelrod. "Nervously," Hamilton recalls, "and rather the way a naturalist might hope to see his first mountain lion in the woods, I had long yearned for and dreaded an encounter with a games theorist. How did they think? What were their dens full of?"[7] This dreaded encounter would lead to a long and fruitful collaboration between Axelrod and Hamilton—a collaboration that would extend the scope of Hamilton's rule to new and uncharted territories.

To examine how the prisoner's dilemma game is used to model the evolution of cooperation between unrelated individuals, imagine that a crime has been committed and the police have arrested two suspects whom they believe are responsible. The suspects are interrogated in separate rooms by the police, and they have two options, or strategies, available to them: to cooperate with each other (that is, stay quiet and tell the police nothing) or to defect (that is, "squeal" and inform the authorities that the other suspect is guilty). Further imagine that absent a confession from either suspect, the police have enough circumstantial evidence to put both suspects in jail for a year, but if both suspects squeal, they both go to jail for three years. Finally, if only one suspect defects, but his partner remains silent, the

Suspect 2

		Cooperate	Defect
Suspect 1	**Cooperate**	R 1-year prison term	S 5-year prison term
	Defect	T 0-year prison term	P 3-year prison term

Figure 9.1. Suspect 1's payoff in the prisoner's dilemma game. T = the temptation to squeal payoff, R = the reward for mutual cooperation payoff, P = the punishment for mutual defection payoff, and S = the "sucker's payoff" for cooperating when one's partner defects.

defector walks away a free man, but his partner to goes to jail for five years.

If we put the jail terms—the payoffs—associated with the possible combinations of cooperation and defection into a single matrix, we have a prisoner's dilemma game. In this game, suspect 1 always receives a higher payoff for defecting than for cooperating. To see why, remember that suspect 2 has two options: cooperate or defect. If suspect 2 cooperates, suspect 1 goes to jail for a year if he cooperates, but walks away a free man if he defects, and so suspect 1 should defect. If suspect 2 defects, suspect 1 goes to jail for five years if he cooperates, but three years if he defects, and so again, suspect 1 should defect. If we then transpose this payoff matrix—that is, if we switch the suspect 2 and suspect 1 labels—we can show that suspect 2 should always defect. When both suspects defect, they each receive a payoff of three years in prison. The "dilemma" is that if both suspects had only cooperated with each other, each would have received a higher payoff (only one year in prison) than they obtained from mutual defection (three years in prison).[8] Each player is tempted to cheat, but both do better if they cooperate.

To determine if cooperative behavior is ever a solution to the prisoner's dilemma game—that is, to examine the evolution of cooperation among unrelated individuals—Axelrod and Hamilton examined the iterated, or repeated, variety of this game. In the iterated prisoner's dilemma game, a pair of players meet each other many, many times, and neither player knows when interactions with a particular partner will end. Rather than being restricted to simple behavioral strategies like "cooperate" and "defect," individuals in an iterated prisoner dilemma can use more complex "if, then" rules, like "cooperate, if your partner cooperates." These sorts of rules were not part of Hamilton's model on altruism and kinship, but would prove invaluable in understanding cooperation among *unrelated* individuals.

To find a solution to the iterated prisoner's dilemma, Axelrod and Hamilton used a combination of computer simulations and analytical mathematics.[9] Both techniques led to the same conclusion.[10] If the probability of meeting a partner again in the future was sufficiently high, then a strategy called *tit for tat (tft)*[11] did very well in the iterated prisoner's dilemma.[12] This is a conditionally cooperative strategy that instructs a player to cooperate on its initial encounter with a new partner and then subsequently copy its partner's previous move—that is, *tft* allows for a limited form of cooperation between unrelated individuals. Axelrod and Hamilton attributed *tft*'s success in the iterated prisoner's dilemma game to three factors associated with this strategy: (1) "niceness"—a player using *tft* is never the first to defect, (2) swift "retaliation"—*tft* players immediately defect on a defecting partner, and (3) "forgiving" behavior—*tft*ers remember only *one move back in time*, and so if its partner cooperates, a *tft* player forgives prior acts of defection.[13]

In a 1981 *Science* paper that was awarded the Newcomb-Cleveland Prize as best publication of the year, Axelrod and Hamilton presented the results of their work on the iterated prisoner's dilemma. "Bob was very pleased to have this [paper] in *Science*," Hamilton recalled, "telling me that this was a rare achievement for a social scientist."[14] Where Hamilton's rule explains cooperation and altruism between relatives, Axelrod and Hamilton's *Science* paper seems to explain the evolution of cooperation and altruism in the absence of genetic relatedness.

There is an interesting wrinkle in the model that Axelrod and Hamilton built. It appears to model cooperation and altruism between unrelated individuals, but individuals in this model are, in fact, relatives of a sort, and so we may be back to Hamilton's rule after all. To see how the *Science* model could really be the long hand of Hamilton's rule at work, we need to take a gene's-eye perspective of the game. If we look at the individuals trapped in Axelrod and Hamilton's prisoner dilemma, they are indeed unrelated—they are not blood relatives in the everyday sense of being brothers or cousins. But if we shift perspectives, and look at this from the point of view of the gene(s) coding for *tft*, we see something quite different.

Image a world full of defectors and *tft*ers. The upshot of the *tft* strategy is that it instructs an individual to only cooperate with others who are clearly cooperators themselves. From the gene's-eye perspective, then, *tft* only doles out cooperation to other individuals that possess the gene for *tft*—in a sense to its *tft* brothers and sisters. And, as Axelrod and Hamilton noted, if *tft* is especially likely to encounter other *tft*ers, this strategy can thrive. Which is to say that *tft* is favored in the prisoner's dilemma game for the exact same reason that altruism is favored in Hamilton's rule—because the gene(s) underlying *tft* can help copies of themselves that reside in other (*tft*) individuals.

The *Science* paper coauthored with Axelrod, along with his original work on kinship and altruism, raised Hamilton to the level of an academic superstar and with that came new opportunities, including that of Royal Society Research Professor—one of the most coveted positions in British academia. This professorship comes with Royal Society funding (for salary and research expenses), and the conditions of the award stipulate that "Holders shall normally devote their whole time to research."[15] In 1984, at the urging of Oxford professor Sir Richard Southwood, Bill Hamilton applied for, and was awarded, a Royal Society Research Professorship at Oxford University. Hamilton had, in fact, dabbled with the idea of returning home a few years earlier and had even flown back to England for an informal interview. But at that point he no great desire to leave the University of Michigan or the pretty little city of Ann Arbor, and he passed on the opportunity. Two things were different when the Royal Society Research Professorship became available. First, the level of violence in the United States had increased to the extent that Ann Arbor was not the place it used to be. Hamilton lamented, "Remembering my own childhood in Kent, and contrasting this with how in Ann Arbor we couldn't even let our young daughter cross the park at the side of our garden to play with friends a half mile away was frustrating."[16]

Second, he felt that political correctness, especially in liberal havens like Ann Arbor, had run amok. For example, Hamilton, though not a smoker himself, was disturbed by "the growing witch hunt directed at tobacco smokers." And then there was the political correctness that was linked to sociobiology. Hamilton held that "women are not as good on average at maths and spatial problems as men" and that "this matter seemed really to need no statistical tests although plenty have been recorded." It was with this mindset that he had written a job reference for a female student of his who was particularly talented at mathematics, in which he noted that she was "exceptionally strong on the theoretical and statistical side and with an ability that was especially remarkable in view of her sex and non-mathematical background." Hamilton notes that he had added the "in view of her sex" phrase to help the student, so that readers would not think she was just good at math as far as women go, but instead good at math period. But, both the search committee reading Hamilton's letter and the student herself were infuriated at the language. Hamilton realized that he was "getting seriously out of touch with America," and that it was time to go home.[17]

Always obsessive about his work, at Oxford Hamilton could at last devote all his time to research. He was always cordial and friendly with departmental colleagues, but when working on some new idea, he would often seal himself in his office for long periods. At such times it was very difficult to pull him away from his desk, but Richard Dawkins took it upon himself to make sure Hamilton attended some of the zoology department research lectures. Dawkins recalls, "It was no use reminding him more than five minutes ahead of time, or sending him written memos. He would simply become reabsorbed in whatever was his current obsession and forget everything else."[18]

Hamilton's 1963 and 1964 papers are far and away the most cited papers in the entire field, and they had a profound impact not only on Dawkins, Emlen, Sherman, and Reeve, but on other leaders in the field of evolution—people like John Krebs, Geoff Parker, Alan Grafen, and Manfred Milinski. His work on altruism and kinship spurred endless dissertation projects and hundred of published papers, both theoretical and empirical. All of this is not to say that Hamilton's rule was the last word with respect to altruism and kinship. Progress on kinship and altruism continues into the present, as Hamilton hoped it would. And debate on these subjects continues as well—just as one might expect for ideas that fundamentally changed the way that people thought about behavior. As we have seen, Hamilton's ideas help explain some, but not all categories of human behavior, and future work needs to better examine the interaction of cultural evolution and genetic evolution (via Hamilton's rule) in shaping human altruism.

There is also continuing debate about the importance of the various components of the equation (r, b, and c) in shaping altruism. E. O. Wilson's recent work on altruism in social insects, for example, suggests that the impact of the b and c terms in Hamilton's rule have be overlooked, despite the fact they may be critically important, particularly when genetic relatedness in social insects is somehow diluted by such things as having multiple queens in a colony. Indeed, recent work suggests social insect colonies may have an average genetic relatedness lower than the $r = 0.75$ that would be the case for groups with a single queen mating with a single male drone. And so, the altruism in social insects may be a result of very high b terms, in conjunction with high, but not as high as first expected values, for genetic relatedness (r). In the end, though, this is not particularly damning for Hamilton's rule.

In the first place, Hamilton's rule is general and was meant to capture altruism in all species, not just the social insects. And, as we have seen, the work from mammals and birds fits quite well with its predictions. Equally important, even when it comes to the social insects, Hamilton never argued that having a genetic relatedness of $r = 0.75$ was all that mattered. While it is true that in the 1964 papers he did put great emphasis on the social insects and the assumed r value of 0.75 between sisters, in totality these papers make it clear that it is the interaction of the r, b, and c terms—not an absolute r value of three-quarters—that determines whether social insect altruism evolves. No doubt very high values of genetic relatedness would make altruism more likely, but they are not a prerequisite for Hamilton's rule to apply to social insect altruism: the only prerequisite being r times b is greater than c.

And so, up until his last days at Oxford, Hamilton would no doubt have been pleased with the impact his model of altruism and kinship had on the field of evolution and behavior. Indeed, despite the fact that Wilson is rightly credited with coining the term "sociobiology" to capture work on the evolution of social behavior, and the fact that Wilson's and Dawkins's books introduced Hamilton's rule to a generation of scientists, when modern sociobiologists are asked to mark the birth of the field, many respond by citing Hamilton's 1963 and 1964 papers on kinship and altruism.

"For me," Hamilton once mused, "the exciting issues were always the biggest I could see, the farthest yet highest of blue mountain tops, which means I was always getting a little tired of my old ones."[19] His great passion during his years at Oxford was understanding the evolution of sexual reproduction, and, in particular, the role that parasites played in such evolution.[20] His basic argument here was that sexual reproduction allowed vast numbers of new genetic combinations for fighting parasitic diseases to be tested by natural selection. As with his work on kinship and altruism, it took time for his ideas on the evolution of sexual reproduction to become widely accepted, but when they did, they became the basis for new research programs all around the world.

Honors eventually followed Hamilton's achievements. He was elected a foreign member of the American Academy of Arts and Sciences in 1978 and a fellow of the Royal Society of London in 1980. He was also awarded the Royal Society's Darwin Medal (1988), the Scientific Medal of the Linnean Society (1989), the Zoological Society of London's Frink Medal (1991), the Kyoto Prize (1993), and the Royal Society of Sweden's Crafoord Award (1993), and he received virtually every other accolade that could possibly be bestowed on an evolutionary biologist.

The power and versatility of Hamilton's rule made evolutionary biologists look to Bill Hamilton as a leader and a man who was expected to spend the last part of his life as an elder statesman in the field. But that was never to be. On March 7, 2000, after weeks in a semicomatose state,[21] sixty-three-year-old Professor William D. Hamilton died as a result of a massive hemorrhage following a very serious bout of malaria. He had caught malaria while he was in the Congo testing the provocative hypothesis that HIV had initially spread from other primates to humans through a botched polio vaccination program undertaken in Africa during the 1950s.[22] He wanted to study the question first-hand, in nature. Before he was able to make much progress, however, he contracted malaria. Although Hamilton responded fairly well to the treatments he received in Africa, he opted to return to England. When his flight landed in England, he felt ill and went to the hospital, but the line to see a doctor was very long, and he decided to return in the morning. The next morning he passed out in a hospital bathroom, lapsed into a semicomatose state, and died five weeks later.

Following his death, articles, tributes, and obituaries quickly appeared in publications ranging from the *New York Times*, the *London Times* and the *Guardian* to *Science* and *Nature*. The *Guardian*'s obituary called Hamilton "the primary theoretical innovator in modern Darwinian biology, responsible for the shape of the subject today"[23] and the *London Times* labeled him "one of the leaders of what has been called 'the second Darwinian revolution.' "[24] The *New York Times* dubbed him "one of the towering figures of modern biology"[25] and the *London Independent* believed that Hamilton was "a good candidate for the title of most distinguished Darwinian since Darwin."[26] Hamilton, however, saw his life quite differently. "I grimace," he wrote, "rub two unrequestedly bushy eyebrows together . . . snort through nostrils that each day more resemble the horse-hair bursts of an Edwardian sofa, and, with my knuckles not yet touching the ground, though nearly, galumph onwards to my next paper."[27]

▲ *Notes* ▲

CHAPTER ONE
A SPECIAL DIFFICULTY THAT MIGHT PROVE FATAL

1. Darwin was well aware of the firestorm his book would create, and fears of how it would be received caused him to delay publishing it for many years. Only when he felt threatened by the imminent publication by Alfred Russell Wallace of similar ideas, did he write in earnest. Even then, he published *On the Origin of Species* only with great reluctance, wanting more time to write a bigger, more comprehensive book. However, it was never to be. "My plans of publication are all changed," he wrote to an old friend upon discovering Wallace's ideas. Darwin despised "the idea of writing for priority," but was equally appalled at the thought that "anyone were to publish my doctrines before me." Darwin to Thomas Eyton, October 4, 1858, and Darwin to Charles Lyell, May 3, 1856, in *The Correspondence of Charles Darwin*, ed. F. Burkhardt and S. Smith, 14 vols. (Cambridge: Cambridge University Press, 1985–2004).

2. H. M. Fraser, *History of Beekeeping in Britain* (London: Bee Research Association, 1958).

3. R. Huish, *A Treatise on the Natural Economy and Practical Management of Bees* (London: Baldwin, Cradock and Joy, 1815).

4. F. R. Prete, "The Conundrum of the Honey Bees," *Journal of the History of Biology* 23 (1990): 272–90.

5. J. Browne, *Charles Darwin: The Power of Place* (New York: Knopf, 2002), 203.

6. C. Darwin, *On the Origin of Species*, 1st ed. (London: J. Murray, 1859), 95, 90, 109.

7. Ibid., 217–18.

8. C. Darwin, *The Voyage of the H.M.S. Beagle* (London: J. Murray, 1845).

9. R. Dawkins, *The Selfish Gene* (Oxford: Oxford University Press, 1976).

10. W. Kirby and W. Spence, *Introduction to Entomology* (London: Longman, Hurst, Rees, Orme and Brown, 1818).

11. R. J. Richards, *Darwin and the Emergence of Evolutionary Theories of Mind and Behavior* (Chicago: University of Chicago Press, 1987), 144.

12. Questions and Experiments notebook 5[a], 9, in P. Barrett, P. J. Guatrye, S. Herbert, D. Kohn, and S. Smith, eds., *Charles Darwin's Notebooks: 1836–1844* (Ithaca, N.Y.: Cornell University Press, 1987).

13. May, 1872, in R. B. Freeman, "Charles Darwin on the Routes of Male Humble Bees," *Bulletin of the British Museum* (Natural History), Historical Series 3: 182.

14. Browne, *Charles Darwin*, 203.

15. Darwin, *Origin*, 236.

16. Ibid., 237. It is worth noting that Darwin was interested in this problem of altruism in all social insects, not just bees; indeed, this quote comes from a discussion relating to ants (another group of social insects).

17. R. Stauffer, ed., *Charles Darwin's Natural Selection, Being the Second Part of His Big Species Book* (Cambridge: Cambridge University Press, 1975), 366.

18. Darwin, *Origin*, 241.

19. Darwin manuscript, June 1848, Darwin's Archive at Cambridge University, as cited in Richards, *Darwin and the Emergence of Evolutionary Theories*, 145, 149.

20. Richards, *Darwin and the Emergence of Evolutionary Theories*, 147.

21. William Youatt, *Cattle: Their Breeds, Management and Disease* (London: Library of Useful Knowledge, 1834).

22. Darwin, *Origin*, 238.

23. In Stauffer, *Charles Darwin's Natural Selection*, 370.

24. Richards, *Darwin and the Emergence of Evolutionary Theory*; also see Prete, "Conundrum of the Honey Bees."

25. Darwin to Huxley, July 3, 1860, in F. Burkhardt, D. Porter, et al., eds., *The Correspondence of Charles Darwin* (Cambridge: Cambridge University Press, 1993).

CHAPTER TWO
DARWIN'S BULLDOG VERSUS THE PRINCE OF EVOLUTION

1. P. Kropotkin, *In Russian and French Prisons* (London: Ward and Downey, 1887).

2. T. H. Huxley, "The Struggle for Existence: A Programme," *Nineteenth Century* 23 (1888): 163–65.

3. P. Kropotkin, *Mutual Aid*, 3rd ed. introduction by George Woodcock (1902; New York: Black Rose Books, 1989), 75.

4. R. Clark, *The Huxleys* (London: Cox, Wyman, 1968).

5. T. H. Huxley, *Autobiography and Selected Essays* (Boston, Houghton Mifflin, 1909), 3.

6. A. Desmond, *Huxley: From Devil's Disciple to Evolution's High Priest* (New York: Addison Wesley, 1994), 3.

7. T. H. Huxley, *Autobiography*, 4.

8. L. Huxley, *Life and Letters of Thomas Henry Huxley* (New York: D. Appleton, 1901), 6.

9. T. H. Huxley, *Autobiography*, 5.

10. Desmond, *Huxley*, 9.

11. Ibid., 11.

12. L. Huxley, *Life and Letters*, 16.

13. Desmond, *Huxley*, 36.

14. Introduction to T. H. Huxley, *Autobiography*, vi.

15. T. H. Huxley, *Autobiography*, 13.

16. Huxley to Darwin, November 23, 1859, in L. Huxley, *Life and Letters*.

17. Quoted in Desmond, *Huxley*, 361.

18. Quoted in ibid., xiii.

19. Huxley to Hooker, December 19, 1860, in L. Huxley, *Life and Letters*.

20. Huxley to Foster, February 14, 1888, in ibid.

21. T. H. Huxley, "Evolution and Ethics: The Romanes Lecture," in J. Huxley, *Touchstone for Ethics, 1893–1943* (New York: Harper and Brothers, 1947), 92.

22. Desmond, *Huxley*, 598.

23. T. H. Huxley, "The Struggle for Existence: A Programme." *Nineteenth Century* 23 (1888): 167, 165, 168, 166.

24. Theologically, Huxley was an agnostic, believing that one simply cannot know whether God exists; in fact, Huxley coined the term "agnostic" (from the Greek "not known"). Even if he had believed in God with any certainty, he no doubt had become quite angry with the Almighty about the time he wrote the gladiator article that so enraged Kropotkin.

25. The official cause of death was pneumonia, but she died at Charcot's Salpetrier Hospital (France) while being treated for her severe mental problems.

26. Draft of Manchester address 1887, as cited in Desmond, *Huxley*, 558n.64.

27. Huxley to Foster, January 8, 1888, in L. Huxley, *Life and Letters*.

28. T. Malthus, *An Essay on the Principle of Population, as It Effects the Future Improvement of Society* (1798; reprint, London: William Pickering,

1986), 1:4. Malthus wrote a much revised second edition in 1803: *An Essay on the Principle of Population, or a View of Its Past and the Present Effects on Human Happiness with an Inquiry into Our Prospects Respecting the Future Removal or Mitigation of the Evils Which It Occasions.*

29. Malthus, *Essay on the Principle of Population*, 5. Malthus, though he recognized that his principle of population applied to animals as well as humans, was clearly concerned with the latter, as evidenced by the subtitle of the revised version of his essay (see note 28).

30. C. Darwin, *Autobiography* (1876), ed. N. Barlow (London: Collins, 1958), 120.

31. C. Darwin, *On the Origin of Species*, 1st ed. (London: J. Murray, 1859), 63.

32. T. H. Huxley, *On the Origin of Species, or The Causes of the Phenomena of Organic Nature* (New York: D. Appleton, 1873), 123–24.

33. Huxley, "The Struggle for Existence," 166.

34. D. Todes, *Darwin without Malthus: The Struggle for Existence in Russian Evolutionary Thought* (New York, Oxford University Press, 1989), 20.

35. N. Riasanovsky, *A History of Russia*, 2nd ed. (New York: Oxford University Press, 1969), 5.

36. Todes, *Darwin without Malthus*, 23. Kovalevskii cited and translated in A. Vucinich, "Russia: Biological Sciences," in T. Glick, ed., *The Comparative Reception of Darwin* (Austin: University of Texas Press, 1972).

37. P. N. Tkachev, "Nauka v poeziia v nauke," *Sochinenie v dvukh tomakh* 1: 398 (as cited in Todes, *Darwin without Malthus*).

38. See Todes, *Darwin without Malthus*, 25, for more on this.

39. K. Timiriazev, "Zemledelie i fizologiia rasteniia," *Sochinenie* 3: 31 (as cited in Todes, *Darwin without Malthus*).

40. P. Kropotkin, "The Scientific Bases of Anarchy," *Nineteenth Century* 21 (1887): 248.

41. N. Danilevskii, ed., *Rossiia i Evropa* (1869; New York: Johnson Reprint Co.), 146 (as cited in Todes, *Darwin without Malthus*).

42. Todes, *Darwin without Malthus*, 29.

43. Letter from Marx to F. Engels, June 18, 1862, Marx-Engels Internet Archive, http://www.marxists.org/archive/marx.

44. See Todes, *Darwin without Malthus*.

45. L. Tolstoy, *What to Do?*, trans. I. Hapgood (New York: Crowell and Co., 1887), 172.

46. R. F. Christian, ed., *Tolstoy's Letters*, vol. 2: 1880–1910 (New York: Charles Scribner's Sons, 1978), 717.

47. This love-hate relationship with Darwin was nicely encapsulated in 1953 by V. V. Dokuchaev, a Russian soil scientist, who praised and criticized Darwin in the same breath, while at the same time mixing religion, science, and politics into one hodgepodge: "The great Darwin, to whom contemporary science is indebted for perhaps 9/10 of its present scope, thought that the world was governed by the Old Testament law: an eye for an eye, a tooth for a tooth. This was a big mistake, a great confusion. One should not blame Darwin for this error.... But Darwin, thank God turns out to have been incorrect. Alongside the cruel Old Testament law of constant struggle we now clearly see the law of cooperation, of love." "Publichnye lektsii po pochvovedeniiu i sel'skomu khoziastvu (1890–1900)," *Sochinenie* 7 (1953): 277 (as cited in Todes, *Darwin without Malthus*).

48. Darwin, *Origin*, 52. While this latter claim is often paraded as an example that struggle need not be taken in the literal sense, it is very clear from the totality of Darwin's arguments that he generally believed the struggle was a real one, a direct struggle between individuals for survival in a world of limited resources. Overpopulation led to this competition, which in turn led to adaptations and even new species.

49. The undisputed founder of the mutual aid school of Russian evolutionary thinking was Karl Fedorovich Kessler (other leaders in this school included N. A. Severtsov, M. A. Menzbir, A. F. Brandt, M. M. Filippov, V. K. Bekhterev, and M. N. Bogdanov). Kessler, who was originally trained as a mathematician, developed an interest in zoology during his years in academia. Eventually, he would become a world-class naturalist and publish an important five-volume series on birds, mammals, and fish. In his travels through an underpopulated Russian landscape, Kessler saw dispersed populations of animals and humans combating a fierce environment. He and others in the Russian school believed that when the environment was harsh, organisms aided one another to fight against the elements. For Kessler and his colleagues, mutual aid was a rule—perhaps the rule—that best explained the evolutionary forces operating in nature.

Kessler's reputation as founder of the Russian mutual aid school is linked to a speech, "On the Law of Mutual Aid," that he presented in December 1879 to the St. Petersberg Society of Naturalists. For Kessler, Darwin's individual versus individual struggle for existence was real, but secondary to the mutual aid that organisms so readily display. He argued, however, that the need to reproduce—the true measure of evolutionary success—required a mutual aid in fish that dwarfed the strength of the individual versus individual conflict associated with foraging. For Kessler, this was just one of many examples that demonstrated what he referred to as "the law of mutual aid."

50. Petr Kropotkin in a letter to Marie Goldsmith, April 7, 1915, Boris Nicolaevsky Collection, Hoover Institution on War, Resolution and Peace, Stanford University, file 7.

51. P. Kropotkin, "The Direct Action of Environment and Evolution," *Nineteenth Century and After* 85 (1919): 89.

52. P. Kropotkin, *Memoirs of a Revolutionist* (1899; rpt., New York: Grove Press, 1968), 5, 8, 28.

53. Ibid., 47.

54. Ibid., 66.

55. Petr's fate in this matter was sealed in rather remarkable circumstances. When he was merely eight years old, an elaborate party was held to celebrate Czar Nicholas's silver jubilee. Naturally, Petr was deemed too young to attend such an affair, but when one of the children of his mother's friend fell ill, he borrowed the beautiful costume that had been designed for the sick child. The costume fit him perfectly, and he went to the party in the other child's place. Then to the delight and amazement of the Kropotkins, Czar Nicholas noticed Petr and ordered him to stand on the platform. After talking with young boy, the Czar announced, "This is the sort of boy you must bring me." Kropotkin, *Memoirs*, 25.

56. Kropotkin, *Memoirs*, 98.

57. Ibid., 146.

58. Ibid., 154.

59. G. Woodcock and I. Avakumovic, *The Anarchist Prince: A Biographical Study of Peter Kropotkin* (London: T. V. Boardman and Co., 1950), 53.

60. Kropotkin, *Mutual Aid*, xxxvii.

61. Woodcock and Avakumovic, *The Anarchist Prince*, 61.

62. Kropotkin, *Memoirs*, 198.

63. Kropotkin, *Mutual Aid*, xxxvi–xxxvii.

64. Ibid.,

65. There is certainly still debate over the exact definitions of altruism. For most people who study social behavior, game theory has become the most common modeling technique used to study altruism and cooperation. Within game theory, both cooperation and altruism are defined as behaviors that entail a cost to the actor and a benefit to

others. In other contexts, cooperation is defined strictly in terms of benefiting all parties involved, and thus entails a benefit, not a cost, assigned to the payoff of the cooperator.

66. Kropotkin, *Mutual Aid*, 10–11.

67. Kropotkin, *Memoirs*, 216.

68. Kropotkin, *Mutual Aid*, 186, 76.

69. Ibid., 79, 83.

70. Ibid., 80.

71. Ibid., 79.

72. Kropotkin, *Memoirs*, 217.

73. P. Kropotkin, "Charles Darwin," *Le Revolte*, April 29, 1882.

74. Kropotkin, *Mutual Aid*, 4.

75. Kropotkin, "The Scientific Bases of Anarchy," 238.

76. Kropotkin, *Mutual Aid*, 186.

77. Kropotkin, "The Scientific Bases of Anarchy," 238–39.

78. Kropotkin, *Memoirs*, 403.

79. S. J. Gould, "Kropotkin Was No Crackpot," *Natural History* 7 (1988): 12–21.

80. Kropotkin in a letter to Marie Goldsmith, August 15, 1909, as cited in Todes, *Darwin without Malthus*, 123, my italics.

81. J. Huxley, *T. H. Huxley's Dairy of the Voyage of the H.M.S.* Rattlesnake (1935; New York: Kraus Reprint, 1972).

82. When Kropotkin first went to Siberia, he was eager to witness the struggle for existence. But even though he tried to slant his early letters to his brother Alexander in that direction, his observations prevented this: As historian of science Daniel Todes puts it, "We can almost hear him thinking aloud attempting to describe the conflicts in Siberian nature in terms of Darwin's emphasis on competition among organisms." But the Siberian wasteland would not lend itself to this description and instead led Kropotkin to the conclusion that mutual aid governed animals' behavior at every turn. Todes, *Darwin without Malthus*, 128.

83. Kropotkin was also the geologist who correctly determined the directional orientation of the major mountain ranges in Asia.

84. Woodcock and Avakumovic, *The Anarchist Prince*, 282. Kropotkin's fame in the United States also had its downside. When President McKinley was shot in 1901 by anarchist Leon Czolgoz, the Chicago newspapers suggested that Kropotkin, who had just recently left the country, might somehow be behind the assassination. This turned out to be sheer nonsense, but a strong crackdown on U.S. anarchists fol-

lowed the assassination (and the subsequent execution of Czolgoz), and Kropotkin was never again allowed entry into the United States after McKinley's death. He did at long last get a chance to return to Russia toward the end of his days. At age seventy-five, Kropotkin was warmly welcomed home by the new Bolshevik government in 1917— so warmly, in fact, that they named a city after him, and Russian Prime Minister Kerensky offered him the position of minister of education, which Kropotkin declined.

85. *Encyclopedia Britannica*, Thomas Henry Huxley.

86. Desmond, *Huxley*, xvii.

87. T. H. Huxley, *Man's Place in Nature*, introduction by Stephen Jay Gould (New York: Random House, 2001), x. Originally published in 1863.

88. P. Marler and D. R. Griffin, "The 1973 Nobel Prize for Physiology or Medicine," *Science* 182 (1973): 464–66; R. A. Hinde, "Nobel Recognition for Ethology," *Nature* 245 (1973): 346.

89. D. Dewsbury, *Studying Animal Behavior: Autobiographies of the Founders* (Chicago: University of Chicago Press, 1985); R. Burkhardt, *Patterns of Behavior: Konrad Lorenz, Niko Tinbergen and the Founding of Ethology* (Chicago: University of Chicago Press, 2005). When some experimental work on animal behavior began to emerge at the start of the twentieth century, it was rough going, with one early investigator lamenting that to keep his work on doves and pigeons afloat, financial circumstances might force him to eat some of his subjects. The researcher was Wallace Craig (see Burkhardt, *Patterns of Behavior*, 15).

90. Kropotkin, *Mutual Aid*, 10.

CHAPTER THREE
THE GREATEST WORD FROM SCIENCE SINCE DARWIN

1. Pronounced AL-lee.

2. G. Mitman, *The State of Nature: Ecology, Community and American Social Thought, 1900–1950* (Chicago: University of Chicago Press, 1992), 48.

3. Warder Clyde Allee Papers (WCAP), Special Collections, University of Chicago, box 26, folder 7.

4. Ibid., box 9, folder 1; and box 25, folder 11.

5. K. P. Schmidt, "Warder Clyde Allee," *Proceedings of the National Academy of Sciences, Biographical Memoirs* 30 (1957): 3–40.

6. Allee was primarily a summer school student at first, teaching high school during the course of the academic year.

7. Mitman, *The State of Nature*, 30.

8. Whitman is credited with giving one of the first formal lectures on animal behavior at the Marine Biological Laboratory in 1898.

9. Mitman, *The State of Nature*, 7.

10. Frank Lillie, "The Department of Biology in Relation to the New Organization," *Daily Maroon*, December 11, 1930.

11. W. C. Allee, "An Experimental Analysis of the Relation between Physiological States and Rheotaxis in Isopoda," *Journal of Experimental Zoology* 13 (1912): 270–344.

12. Obituary for Majorie Hill Allee, WCAP, box 25, folder 3.

13. M. H. Allee, *Jane's Island* (New York: Houghton Mifflin, 1931).

14. W. C. Allee and M. H. Allee, *Jungle Island* (New York: Rand McNally, 1925).

15. Schmidt, "Warder Clyde Allee," 7.

16. Mitman, *The State of Nature*, 46 n.103.

17. W. C. Allee, "Animal Aggregations," *Quarterly Review of Biology* 2 (1927): 367–98.

18. Ibid., 377, 379.

19. W. C. Allee, *Animal Aggregations: A Study in General Sociology* (Chicago: University of Chicago Press, 1927). A second edition appeared in 1931.

20. Ibid., 204, and Allee, "Animal Aggregations," 380–81.

21. Allee, "Animal Aggregations," 379.

22. W. C. Allee, *Cooperation among Animals with Human Implications* (New York: Henry Schuman, 1951), 26.

23. A. Popovici-Baznosanu, "L'influence de quelques sur l'accroissement des Gastropodes d'eau douce," *Archives de Zoologie Experimentale et Générale* 60 (1921): 501–21.

24. Allee, *Cooperation among Animals*.

25. T. B. Robertson, "Experimental Studies on Cellular Reproduction: II. The Influence of Mutual Contiguity upon Reproductive Rate in Infusoria," *Biochemical Journal* 15 (1921): 612–19.

26. Allee, "Animal Aggregations," 385.

27. V. L. Kellogg, *Headquarters Nights* (Boston: Atlantic Monthly Press, 1917).

28. Watson to Allee, December 29, 1949, WCAP, box 21, folder 4; and Allee to Watson, December 31, 1949, WCAP, box 21, folder 4.

29. *Springfield News-Record*, February 9, 1917.

30. Quoted in E. Banks, "Warder Clyde Allee and the Chicago School of Animal Behavior," *Journal of the History of Behavioral Science* 21 (1985): 351.

31. Only the draft version of this article is in Allee's files, but it appears to have been aimed at a Quaker audience.

32. W. C. Allee, "Reexamination of One Fundamental Doctrine in the Light of Modern Knowledge," WCAP, box 26, folder 3.

33. W. C. Allee, "Concerning the Biology of War," June 1940, WCAP, box 2, folder 7.

34. Allee, "Animal Aggregations," 387, 379.

35. For background information on Sewall Wright, see the American Philosophical Society at http://www.amphilsoc.org/library/mole/w/wrights.htm.

36. W. Provine, *The Origins of Theoretical Population Genetics* (Chicago: University of Chicago Press, 1971); and *Sewall Wright and Evolutionary Biology* (Chicago: University of Chicago Press, 1986).

37. S. Wright, "Tempo and Mode in Evolution: A Critical Review," *Ecology* 26 (1945): 415–19.

38. D. S. Wilson, "The Group Selection Controversy: History and Current Status," *Annual Review of Ecology & Systematics* 14 (1983): 159–87.

39. D. Wilson, "Structured Demes and the Evolution of Group-Advantageous Traits," *American Naturalist* 111 (1977): 157–85; D. S. Wilson, *The Natural Selection of Populations and Communities* (Menlo Park, Calif.: Benjamin Cummings, 1980); R. Dawkins, *The Selfish Gene* (Oxford: Oxford University Press, 1976); R. Dawkins, "Twelve Misunderstandings of Kin Selection," *Zeitscrift fur Tierpsychologie* 51 (1979): 184–200; E. Sober and D. S. Wilson, *Unto Others* (Cambridge, Mass.: Harvard University Press, 1998).

40. D. S. Wilson, *The Natural Selection of Populations and Communities* (Menlo Park, Calif.: Benjamin Cummings, 1980); L. A. Dugatkin and H. K. Reeve, "Behavioral Ecology and Levels of Selection: Dissolving the Group Selection Controversy," *Advances in the Study of Behaviour* 23 (1994): 101–33; D. C. Queller, "Group Selection and Kin Selection," *Trends in Ecology & Evolution* 6 (1991): 64.

41. Modern group selectionists would argue that the family group is just one type of group, but kinship did not play a role in the group selection that Wright and Allee were discussing. Wright's model was attractive to researchers studying ecology and behavior in Chicago's zoology department, not just because its creator was both brilliant and

one of their own, but because of the emphasis that his model placed on the *population*. Allee and his physiological ecologist peers were trained to think of the world not so much in terms of individuals as of populations. Indeed, such a population-centered approach to ecology and behavior became their calling card. Allee argued that if ecological studies showed one thing, it was that any "general biological theory," including a theory of human behavior, would revolve around the notion of a population. In Allee's eyes, Wright's model laid the mathematical framework for such a claim.

42. S. Wright, "Size of Population and Breeding Structure in Relation to Evolution," *Science* 87 (1938): 430–31; Wright, "Tempo and Mode in Evolution."

43. Allee to Julian Huxley, April 4, 1939, WCAP, box 18, folder 15.

44. Mitman, *The State of Nature*, 13.

45. Allee, "Animal Aggregations."

46. Allee, *Animal Aggregations*, 338.

47. Allee, *Cooperation among Animals*, 28.

48. Ibid., 26.

49. Allee, *Animal Aggregations*, 339–40.

50. Allee, "Animal Aggregations," 390.

51. Quoted in ibid., 391.

52. Mitman, *The State of Nature*, 81–84. While most anthropologists saw their views on kinship and cooperation as antithetical to Allee's, there were exceptions, one of whom was his friend Ralph Linton, author of *The Study of Man*. Linton's theories seemed eerily reminiscent of Allee's ideas on family life and cooperation, just cast in anthropological terms. It is possible that Allee and Linton met at the Hyde Park gatherings of the Society of Friends, a Quaker group with which Allee was associated for decades. R. Linton, *The Study of Man: An Introduction* (New York: Appleton, 1936); A. Linton and C. Wagley, *Ralph Linton* (New York: Columbia University Press, 1971).

53. Funded indirectly by the Rockefeller Foundation.

54. W. C. Allee, "Where Angels Fear to Tread: A Contribution from General Sociology to Human Ethics," *Science* 97 (1943): 517–25.

55. W. C. Allee, "Cooperation among Animals," *University of Chicago Magazine*, June 1928, 424.

56. Allee, *Animal Aggregations*, 362.

57. Mitman, *The State of Nature*, 84.

58. Allee, "Where Angels Fear to Tread," 524.

59. Newman to Allee, November 17, 1953, WCAP, box 9, folder 4.

60. "Concerning Biology and Biologists" WCAP, box 9, folder 4.
61. Ibid.
62. Allee, *Animal Aggregations*, 355.
63. Allee, *Cooperation among Animals*, 213.
64. October 26, 1946, WCAP, box 2, folder 6.
65. Allee to Phillips, February 13, 1933, WCAP, box 21, folder 3.
66. Mitman, *The State of Nature*, 139.
67. Griggs to Allee, January 31 and March 31, 1947, WCAP, box 2, folder 6.
68. WCAP, box 26, folder 7.

CHAPTER FOUR
J.B.S.: THE LAST MAN WHO MIGHT KNOW
ALL THERE WAS TO BE KNOWN

1. J.B.S. Haldane and J. Huxley, *Animal Biology* (Oxford: Claredon, 1927).
2. J.B.S. Haldane, *Possible Worlds* (New York: Harper Brothers, 1928), 298.
3. For a fascinating history of Haldane's "inordinate fondness" quote, see S. J. Gould, "A Special Fondness for Beetles," *Natural History*, January 1993, 4–11. Two other published versions of this story have Haldane noting: "the creator would appear as endowed with a passion for stars, on the one hand, and beetles on the other, for the simple reason that there are nearly 300,000 species of beetles known" (J.B.S. Haldane, *What Is Life?* [New York: Boni and Gaer, 1947], 239); and "the creator, if he exists, has a special preference for beetles, and so we might be more likely to meet them than any other type of animal on a planet that would support life" (A. Clarke, "Haldane and Space," in *Haldane and Modern Biology*, ed. K. Dronamraju [Baltimore: Johns Hopkins University Press, 1968], 244).
4. K. Dronamraju, *Haldane: The Life and Work of J.B.S. Haldane with Special Reference to India* (Aberdeen: Aberdeen University Press, 1985), xv.
5. Clarke, "Haldane and Space," 248.
6. Quoted in ibid., 243–48.
7. N. Pirie, "John Burdon Sanderson Haldane: 1892–1964," *Biographical Memoirs of Fellows of the Royal Society* 12 (1966): 219–49. Included in Haldane's popular writings were almost 350 articles he wrote during

his tenure as science correspondent for the communist paper the *Daily Worker*.

8. Pirie, "John Burdon Sanderson Haldane," 219, quoted from J.B.S. Haldane, "The Scientific Work of J. S. Haldane," *Penguin Science Survey* 2 (1961): 11.

9. Haldane as quoted in R. Clark, *JBS: The Life and Work of J.B.S. Haldane* (London: Hodder and Stoughton, 1968), 13.

10. Ibid., 26.

11. Ibid., 21.

12. Haldane and Huxley, *Animal Biology*.

13. As quoted in Clark, *JBS*, 22.

14. For more on Goodrich, see M. Ruse, *From Monads to Man* (Cambridge, Mass.: Harvard University Press, 1996).

15. Haldane as quoted in Clark, *JBS*, 27.

16. Clark, *JBS*, 37.

17. Pirie, "John Burdon Sanderson Haldane," 219; J.B.S. quote from J.B.S. Haldane, *The Inequality of Man and Other Essays* (London: Chatto and Windus, 1941).

18. Haldane as quoted in Clark, *JBS*, 39.

19. J.B.S. Haldane, A. D. Sprunt, and M. N. Haldane, "Reduplication in Mice," *Journal of Genetics* 5 (1915): 133–35.

20. Usually this occurs when the genes in question reside near one another on a chromosome (the linear strands of DNA that reside in a cell's nucleus) and do not segregate independently during cell division. One example of genetic linkage involves the anthers and stigma of the primrose, *Primula vulgaris*. In the "pin" form of the primrose, the stigma (which receives pollen) sits above the anthers (which produce pollen), while in the "thrum" form, the anthers sit above the stigma. Pin and thrum varieties differ at two very closely linked locations (known as loci) on a chromosome. One of these loci (the G locus) controls stigma height, while the other (the A locus) controls anther height. Each gene has two variants, or alleles; in this case we have G, g, A, and a. Genetic linkage occurs in this system, as G is almost always found with A, and g with a; the pin form always possesses the ga/ga genotype, while the genotype for thrum is GA/ga. The G and A loci are linked and inherited as if they were one unit—a supergene. E. B. Ford, *Ecological Genetics* (London: Chapman and Hall, 1971); D. Futuyma, *Evolutionary Biology* (Sunderland, Mass.: Sinauer Press, 1986).

21. John Maynard Smith, *Haldane, J.B.S. Science and Life: Essays of a Rationalist* (London: Humanist Library, 1968), ix.

22. Haldane as quoted in Clark, *JBS*, 65, 63.

23. Norman Pirie as quoted in ibid., 66.

24. Haldane as quoted in ibid., 79.

25. Ibid., 80.

26. Bateson quoted in Clark, *JBS*, 79.

27. Clark, *JBS*, 80–82.

28. Much of the work in these papers is summarized for the lay reader in J.B.S. Haldane, *The Causes of Evolution* (London: Longmans Green, 1932).

29. Haldane, *Causes of Evolution*.

30. E. Mayr, "Haldane's *Causes of Evolution* after 60 Years," *American Naturalist* 67 (1992): 175–86; J. Maynard Smith, "In Haldane's Footsteps," 347–56 in *Studying Animal Behavior: Autobiographies of the Founders* ed. D. Dewsbury (Chicago: University of Chicago Press, 1985).

31. Haldane, "A Mathematical Theory of Natural and Artificial Selection, Part II" *Proceedings of the Cambridge Philosophical Society* 1 (1924): 158.

32. Haldane, *Possible Worlds*, 35.

33. Ibid.

34. For more on Haldane's view of the power of science, see Ruse, *From Monads to Man*, as well as Haldane, *Possible Worlds*; J.B.S. Haldane, *Science Advances* (New York: Macmillan, 1948); and Haldane, *Science and Life*.

35. Clarke, "Haldane and Space," 138.

36. J.B.S. Haldane, "Population Genetics," *New Biology* 18 (1955): 34–51.

37. Haldane, *Causes of Evolution*, 1.

38. Ibid., 71.

39. Ibid., 119.

40. In the appendix to *Causes*, Haldane does in fact go beyond a verbal model of altruism, providing the mathematically trained reader with some equations. The model, however, does not consider the effects of kinship per se, and rather generally demonstrates that altruism will be rare in large populations, and by implication, rare in nonkin-based populations.

41. Haldane, "Population Genetics," 44.

42. After Haldane, the closest anyone would come to a gene-counting approach to kinship and social behavior would be a two-paragraph summary in an obscure and lengthy volume on population genetics.

NOTES TO CHAPTER FOUR

This more than nine hundred-page volume, *The Genetics of Populations*, was a monograph published by an animal breeder/geneticist named Jay L. Lush. Lush, completely in passing, notes the following:

> The competition and selection between families . . . could make selection favor any genes which tend to cause the possessor to sacrifice himself . . . provided the sacrifice promotes the biological welfare of his relatives (some of whom will have some of the genes he has) enough to more than compensate for the genes lost in his own self-sacrifice. . . . The balance here is intricate, depending on how extreme the self-sacrifice, at what stage in the life cycle it occurs, how much the this increases the biological success of the benefactor's relatives, how closely related they are, etc. The balance depends primarily on whether the greater biological success of the relatives will multiply the genes for altruism more than enough to compensate for the loss of those genes in the individual which sacrificed itself."

After this brief treatment, Lush never again touches on the subject. J. Lush, *The Genetics of Populations*, Special Report 94 (Ames: Iowa State University, College of Agriculture, Iowa Agriculture and Home Economics Experiment Station, 1948).

43. Haldane, *Causes of Evolution*, 71, 120.

44. Ibid., 18.

45. R. A. Fisher, "On the Dominance Ratio," *Proceedings of the Royal Society of Edinburgh* 42 (1922): 321–22.

46. R. A. Fisher, *The Genetical Theory of Natural Selection* (Oxford: Claredon Press, 1930).

47. R. A. Fisher, "The Correlation between Relatives on the Supposition of Mendelian Inheritance," *Transactions of the Royal Society of Edinburgh* 52 (1918): 399–433.

48. His original table is actually a seven-by-five matrix with thirty-five entries, but for ease of presentation, I have deleted one column, reducing this to a six-by-five matrix.

49. W. Provine, *The Origins of Theoretical Population Genetics* (Chicago: University of Chicago Press, 1971), 140.

50. T. Eisner and D. Aneshansley, "Spray aiming in the bombardier beetle: photographic evidence," *Proceedings of the National Academy of Sciences* 96 (1999): 9705–9.

51. Fisher, *Genetical Theory of Natural Selection*, 158.

52. Ibid., 159.

53. Ibid.

54. Ibid., 163.

55. S. Wright, "Systems of Mating, I: The Biometric Relation between Parent and Offspring," *Genetics* 6 (1921): 111–23.

56. S. Wright, "Coefficients of Inbreeding and Relationship," *American Naturalist* 56 (1922): 330–38.

57. This idea was suggested by an anonymous review of an early version of this book.

58. For example, Fisher, "Correlation between Relatives"; Wright, "Systems of Mating"; and Haldane, "A Mathematical Theory of Natural and Artificial Selection, Part II," *Transactions of the Cambridge Philosophical Society* 23 (1924).

59. Author's interview with Thomas Seeley. Ithaca, New York, January 31, 2003. A similar opinion was expressed by Harvard's Naomi Pierce in an interview with me on February 24, 2003.

60. Author's interview with Kern Reeve. Ithaca, New York, January 31, 2003.

61. J. Maynard Smith, "J.B.S. Haldane," *Oxford Surveys in Evolutionary Biology* 4 (1987): 9.

62. This idea was also suggested by an anonymous review of an early version of this book.

63. Personal communication from Paul Ewald.

Chapter Five
Hamilton's Rule

1. W. D. Hamilton, *Narrow Roads of Gene Land: The Collected Papers of W. D. Hamilton*, vol. 1 (Oxford: W. H. Freeman, 1996), viii–ix.

2. The Callender-Hamilton bridge design remains on the market to this day. For more on Archibald Hamilton's work, see A. M. Hamilton, *Road through Kurdistan: The Narrative of an Engineer in Iraq* (New York: AMS Press, 1937).

3. Author's interview with Mary Bliss, Bill Hamilton's sister, Sevenoaks, Kent, England, March 9, 2003.

4. A. Grafen, "William David Hamilton," *Biographical Memoirs of the Fellows of the Royal Society of London* 50 (2004): 109–32.

5. M. Bliss, "Professor William Donald Hamilton: Memoirs of My Brother's Youth," speech read at the memorial ceremony for W. D. Hamilton in the chapel of the New College, London, July 1, 2000.

6. Hamilton's lecture when accepting the prestigious Kyoto Prize, British Museum, Hamilton Archive.

7. For an excellent review of the life and work of Francis Galton, see M. G. Bulmer, *Francis Galton: Pioneer of Heredity and Biometry* (Baltimore: Johns Hopkins University Press, 2003).

8. Hamilton in *The Secret of the Clouds* videotape. British Library. Originally aired on "The Edge Show."

9. Kyoto Prize lecture.

10. Bliss, "Professor William Donald Hamilton."

11. E-mail message from Mary Bliss to author, April 18, 2003.

12. British Library, BL:z1X55 17.5, June 3 and 6, July 18, 1956. This journal runs from June 3, 1956, to May 27, 1957, and is 146 pages in length.

13. Hamilton, *Narrow Roads*, 1:21.

14. W. D. Hamilton, *Narrow Roads of Gene Land: The Collected Papers of W. D. Hamilton*, vol. 2 (Oxford: Oxford University Press, 2001), 543, 542.

15. From Hamilton's endorsement of the 1999 Oxford University edition of *The Genetical Theory of Natural Selection.*

16. Despite the fact that Fisher's book had a profound effect on him while he studied at Cambridge, Hamilton notes that he did not have Fisher, or for that matter, Haldane's verbal model for the evolution of altruism "consciously in mind" during the early development of his own ideas on this subject. "One reads and forgets," Hamilton recounted. "Hints not understood probably leave their traces, but one has to return to the topic in a better state of preparation and to re-read before such throwaway items become meaningful." *Narrow Roads*, 1:22.

17. Who would one day be knighted.

18. Some fifteen years later, Leach would become one of the most severe critics of E. O. Wilson's *Sociobiology: The New Synthesis* (Cambridge, Mass.: Harvard University Press, 1975).

19. Hamilton, *Narrow Roads*, 1:24.

20. Bill Hamilton to Mary Bliss, November 11, 1959, cited in Bliss, "Professor William David Hamilton."

21. In so doing, Hamilton was following in the footsteps of Ronald Fisher, who was a schoolteacher early in his career.

22. Hamilton, *Narrow Roads*, 1:24.

23. Bliss, "Professor William Donald Hamilton," citing a February 8, 1960, letter from Bill Hamilton.

24. Hamilton, *Narrow Roads*, 1:11.

25. Ibid., 1:25, ix.

26. Ibid., 1:25. John Hajnal, who came to supervise Hamilton at the LSE when his ideas became "too genetical" for Norman Carrier, recalls

a young man who "was worried that the work wouldn't be recognized. Might not even be published . . ." (*The Secret of the Clouds* videotape).

27. In fact, during his early days in graduate school, Hamilton once again considered an alternative career path—this time, as a carpenter. In the end, though, his obsession with evolution, kinship, and altruism would overpower any ideas about other careers.

28. Hamilton, *Narrow Roads*, 2:543.

29. Ibid., 2:481.

30. "The Evolution of Altruistic Behaviour" was initially submitted to *Nature*, the premier English science journal. Hamilton may at times have believed that he was a crank, but he clearly recognized the potential importance of his work in choosing the journal that Thomas Henry Huxley and the X-Club initiated some one hundred years earlier. The editors at *Nature* saw things differently, and Hamilton recalls receiving "the editor's decision almost [immediately] by return post. . . . He had no space for my manuscript and suggested that, given its specialized topic, it might be more appropriate to a psychological or sociological journal" (Hamilton, *Narrow Roads*, 1:3). From the tone of the editor's comments, Hamilton believed it may have been a "tactical mistake" to list his address as the department of sociology at LSE. Perhaps, he thought, *Nature* might have looked at the manuscript in a different, perhaps more serious, light if he used his alternative address at the Galton Laboratory of Genetics at University College. Nonetheless, in a show of loyalty to those at LSE, he once again used that address alone when submitting the manuscript to the *American Naturalist*.

31. W. D. Hamilton, "The Evolution of Altruistic Behavior," *American Naturalist* 97 (1963): 354.

32. Although the argument can get technical, Hamilton later claimed that Wright's coefficient of relatedness was not the perfect variable to use in his model. He would come to recognize, with the help of Richard Michod, that his equation "requires, in principle, a regression coefficient of genotype of recipient on genotype of altruist, whereas Wright's coefficient is the corresponding correlation coefficient." R. E. Michod and W. D. Hamilton, "Coefficients of Relatedness in Sociobiology," *Nature* 288 (1980): 694–97.

33. In his 1963 paper, Hamilton folded the costs and benefits into a single variable, which he labeled k—where k is simply the benefits divided by the costs, or b/c.

34. I thank Oxford University's Alan Grafen for emphasizing the importance of economics in Hamilton's model.

35. This is a slightly different way of calculating things than Haldane adopted in the *New Biology* paper.

36. For instance, it applied only to haploid organisms, and to the case where the altruistic allele was rare.

37. Hamilton, *Narrow Roads*, 1:230.

38. Ibid., 1:3.

39. J. Maynard Smith, "J.B.S. Haldane," *Oxford Surveys in Evolutionary Biology* 4 (1987): 9.

40. J. Maynard Smith, "Group Selection and Kin selection," *Nature* 201 (1964): 1,145–46.

41. V. C. Wynne-Edwards, *Animal Dispersion in Relation to Social Behavior* (Edinburgh: Oliver and Boyd, 1962).

42. Maynard Smith, "Group Selection," 1,145.

43. W. D. Hamilton, "The Genetical Evolution of Social Behaviour, I and II," *Journal of Theoretical Biology* 7 (1964): 1–52.

44. In "Death of an Altruist," Maynard Smith told Jim Schwartz, "I seem to have this fate of getting ideas from other people's manuscripts when I referee them." With respect to the Hamilton *Journal of Theoretical Biology* papers, Maynard Smith noted, "I wasn't trying to steal his idea, or I don't think I was, so it wasn't conscious." J. Schwartz, "Death of an Altruist," *Lingua Franca* 10 (2000): 51–61.

45. Author's interview with Naomi Pierce, Harvard University, Cambridge, Mass., February 25, 2003.

46. Hamilton, "The Genetical Evolution of Social Behaviour," 2.

47. In this paper, Hamilton used his address at University College rather than the London School of Economics: W. D. Hamilton, The Galton Laboratory, University College, London, WC2.

48. Hamilton, "The Genetical Evolution of Social Behaviour," 8.

49. It could hardly have been otherwise; recall that the *American Naturalist* paper (1963), though published before the *Journal of Theoretical Biology* papers (1964), was in fact written after Hamilton had derived the models in the 1964 paper.

50. Hamilton, "The Genetical Evolution of Social Behaviour," 19.

51. Removal of ectoparasites is the most obvious benefit accrued from grooming. Although there appears to be agreement on the importance of ectoparasite removal in the initiation of social grooming, Dunbar's comparative study of forty-four free-living species of primates suggests that grooming's current function is more social than hygienic.

R. Dunbar, "Functional significance of social grooming in primates," *Folia Primatologica* 57 (1991): 121–31. Dunbar hypothesized that if ectoparasite and hygiene-oriented aspects of grooming were important, then we might expect a positive correlation between body size and amount of time spent grooming. Conversely, if social aspects of grooming were paramount, a positive correlation between time spent grooming and group size should be apparent. The latter correlation was uncovered, but the former was not, suggesting that social aspects of grooming are of greater importance (at least in Old World monkeys). Other benefits associated with grooming include "tension reduction"; "coalition formation"; the exchange of grooming for other currencies such as "friendship"; reduced aggression from other individuals, particularly dominant group members; access to scarce resources such as water or food; access to a dominant's dominant "friends"; entrance into new groups; aid in chasing potential predators away; and future association with individuals with "special skills" that others do not possess.

52. Work over the past forty years has generally supported Hamilton's claim that the donor and recipient in grooming bouts will often be related, but other factors, such as the degree of reciprocity between donor and recipient, are often in play as well.

53. Hamilton, "The Genetical Evolution of Social Behavior," 21.

54. Ibid.

55. The high relatedness between sister workers in hymenopterans can be a fragile commodity: the r value of 0.75 that sisters share is contingent on the queen of their hive mating with just one male. If the queen mates with two drones, and each male fathers half the young, then worker relatedness is cut in half, to $r = 0.375$. When the queen mates with more males, r decreases to an even greater extent. Hamilton warns the reader of his 1964 papers that "multiple insemination will greatly weaken the tendency to evolve worker-like altruism," and that when a queen mates with more than two males, this "should prevent its [altruism's] incipience altogether." It appears that Hamilton was overstating the devastating effects of multiple matings on altruism in social insects. Recent work suggests that altruism is common in species where queens mate with many males. In these species, blood relatedness, in conjunction with other factors, may still favor the evolution of cooperation. In any case, no matter how many mates a queen has, hymenopteran sisters are always more related to each other than are hymenopteran brothers and sisters, and so inclusive fitness theory

predicts that if altruism is to evolve to any degree, it should almost always be among hymenopteran sisters.

56. For example, see R. Gadagkar, "Evolution of Eusociality: The Advantages of Assured Fitness Returns," *Philosophical Transactions of the Royal Society of London* 329 (1990): 17–25; D. C. Queller, "The Evolution of Eusociality: Reproductive Head Starts of Workers," *Proceedings of the National Academy of Sciences* 86 (1989): 3,224–26; E. O. Wilson and B. Holldobler, "The Rise of the Ants: A Phylogentic and Ecological Explanation," *Proceedings of the National Academy of Sciences* 102 (2005): 7,411–14.

57. These journals are currently housed in the British Library. The first (British Library: Z1X55 17.1) runs from September 12, 1963, through November 11, 1963. The second (BL: Z1X55 17.14) runs from November 12, 1963, through June 1, 1964. The third starts on June 1, 1964, and ends on July 26, 1964. The address on the journals is Sr. W. D. Hamilton, Faculdade de Filosofia Ciencias E Letra De Rio Claro, Rio Clare, Estado De São Paulo, Brasil.

58. Originally published in Japanese in W. D. Hamilton, "My Intended Burial and Why," *Insectarium* 28. Reprinted in English in W. D. Hamilton, "My Intended Burial and Why," *Ethology Ecology and Evolution* 12 (1991): 111–22.

CHAPTER SIX
THE PRICE OF KINSHIP

1. Hamilton, *Narrow Roads*, 1:85, 86.

2. Ibid., 1:171.

3. Now known as the International Institute of Entomology.

4. W. D. Hamilton, "The Moulding of Senescence by Natural Selection," *Journal of Theoretical Biology* 12 (1966): 12–45. Although one reviewer of the initial manuscript referred to the model as "verbal diarrhea," this was clearly a minority position, as the Hamilton model of aging is today one of the most important ever constructed on the subject.

5. W. D. Hamilton, "Extraordinary Sex Ratios," *Science* 156 (1967): 477–87.

6. W. D. Hamilton, "Geometry of the Selfish Herd," *Journal of Theoretical Biology* 31 (1971): 295–311.

7. W. D. Hamilton, "Altruism and Related Phenomena, Mainly in Social Insects," *Annual Review of Ecology & Systematics* 3 (1972):

192–232; W. D. Hamilton, "Selfish and Spiteful Behaviour in an Evolutionary Model," *Nature* 228 (1972): 1,218–19.

8. J. Schwartz, "Death of an Altruist," *Linga Franca* 10 (2000): 52.

9. For research on uranium analysis as part of the Manhattan Project. British Library, George Price file, BL21X 102_13.

10. G. R. Price, "Science and the Supernatural," *Science* 122 (1955): 359–67; G. R. Price, "Where Is the Definitive Experiment?" *Science* 123 (1956): 17–18. In these papers, Price argued that not a single instance of purported extrasensory perception meets even the most liberal definitions of a scientifically demonstrable phenomenon. This article was picked up by newspapers all over world, including a front-page story in the *New York Times*. British Library, George Price file, BL21X 102_13.

11. British Museum, George Price file, BL21X 102_13.

12. Price most likely came across Hamilton's model in the Senate House Library. Hamilton, *Narrow Roads*, 1:.

13. Hamilton, *Narrow Roads*, 1:173.

14. Because natural selection works on relative fitness, spite can, in principle, evolve, as the relative fitness of the spiteful individual, though decreased by its own action, is still greater than that of the recipient of the spiteful act. The problem that all models of spite come up against is that individuals who are not spiteful at all should have higher fitnesses than either the initiator or recipient in a spiteful encounter.

15. E. C. Waltz, "Reciprocal Altruism and Spite in Gulls: A comment," *American Naturalist* 118 (1981): 588–92. See also R. Pierotti, "Spite, Altruism, and Semantics—A Reply to Waltz," *American Naturalist* 119 (1982): 116–20; K. R. Foster, T. Wenseleers, and F.L.W. Ratnieks, "Spite: Hamilton's Unproven Theory," *Annales Zoologici Fennici* 38 (2001): 229–38; L. Keller, M. Milinski, M. Frischknecht, N. Perrin, H. Richner, and F. Tripet, "Spiteful Animals Still to Be Discovered," *Trends in Ecology & Evolution* 9 (1994): 103; R. Gadagkar, "Spiteful Animals Still to Be Discovered—Reply," *Trends in Ecology & Evolution* 9 (1994): 103; R. Gadagkar, "Can animals Be Spiteful?" *Trends in Ecology & Evolution* 8 (1993): 232–34.

16. Hamilton, *Narrow Roads*, 1:175.

17. W. D. Hamilton, "Selfish and Spiteful Behaviour in an Evolutionary Model," *Nature* 228 (1970): 1,218–19; G. R. Price, "Selection and Covariance," *Nature* 227 (1970): 520–21.

18. Schwartz, "Death of an Altruist," 52, 56.

19. Hamilton, *Narrow Roads*, 1:322–23.

20. Schwartz, "Death of an Altruist," 56.

21. Indeed, Cedric Smith spent considerable time writing grant proposals whose function was to support George Price's work. British Library, George Price file, BL21X 102_13.

22. Interview with John Maynard Smith. This interview can be found in L. A. Dugatkin, *Principles of Animal Behavior* (New York: W. W. Norton, 2004), 502–3.

23. Schwartz, "Death of an Altruist," p. 58.

24. Ibid., 60. In the long run, Price realized he needed to shield the others in the Galton Lab from this sort of thing, so in March 1974, he took a room in the home of an elderly woman he knew and began work as a night janitor. Then in August he quit his job and moved into a squatter's tenement at Tolmer's Square.

25. Price had apparently attempted suicide once before, but precisely when is not known.

26. Hamilton, *Narrow Roads*, 1:174, 323–24.

CHAPTER SEVEN
SPREADING THE WORD

1. R. Dawkins, *The Selfish Gene* (Oxford: Oxford University Press, 1976). E. O. Wilson, *Sociobiology: The New Synthesis* (Cambridge, Mass.: Harvard University Press, 1975). A third book, Jerram L. Brown's *The Evolution of Behavior*, which was published at about the same time as *The Selfish Gene* and *Sociobiology*, also discussed Hamilton's ideas in depth. This book was in many ways both more detailed and more "academic" in nature. While often adopted in graduate classes of the day, *Evolution of Behavior* did not have the wide-ranging effect of either Dawkins's or Wilson's books. J. L. Brown, *The Evolution of Behavior* (New York: W. W. Norton, 1975).

2. Author's interview with Richard Dawkins, Royal Society of London, London, March 10, 2003.

3. G. Williams, *Adaptation and Natural Selection* (Princeton, N.J.: Princeton University Press, 1966), 5.

4. Interview with Dawkins, March 10, 2003.

5. Dawkins, *The Selfish Gene*, 97. In a later, more detailed analysis, Dawkins found that it was not until 1974 that citations to Hamilton's inclusive fitness work began to grow. Thenceforth, "like a gathering epidemic," references to Hamilton's altruism models increased steadily.

6. For example, in 1997, Gould and Dawkins reviewed each other's books in back-to-back papers appearing in *Evolution*: S. J. Gould, "Self-help for a Hedgehog Stuck on a Molehill (A Review of Richard Dawkins's *Climbing Mount Improbable*)," *Evolution* 51 (1997): 1,020–23; R. Dawkins, "Human Chauvinism (A Review Stephen Jay Gould's *Full House*)," *Evolution* 51 (1997): 1,015–20.

7. B. Rensberger, "Updating Darwin on Behavior," *New York Times*, May 28, 1975, 85. For more on the debate surrounding *Sociobiology*, see U. Segerstrale, *Defenders of the Truth: The Battle for Science in the Sociobiology Debate and Beyond* (New York: Oxford University Press, 2000); J. Alcock, *The Triumph of Sociobiology* (New York: Oxford University Press, 2001).

8. Gould was senior author of a November 13, 1975, letter entitled "Against Sociobiology," *New York Review of Books*; E. O. Wilson, "For Sociobiology," *New York Review of Books*, December 11, 1975; S. J. Gould and R. Lewontin, "The Spandrels of San Marcos and the Panglossian Paradigm: A Critique of the Adaptationist Programme," *Proceedings of the Royal Society of London* (ser. B) 205 (1979): 581–98.

9. J. Schwartz, "Oh My Darwin," *Linga Franca* (November 1999).

10. E. O. Wilson, *Naturalist* (New York: Warner Books, 1994), 319, 320.

11. Ibid., 119, 415, 417–18.

12. Although he agrees that when the benefit of altruistic behavior discounted by the degree of relationship exceeds the cost, then altruism evolves, he is also in the midst of developing a new model that suggests "a supervening force," which he calls "ecological selection," could "virtually erase any effect of kin selection" in social insects. Author's interview with E. O. Wilson, Harvard University, Cambridge, Mass., February 26, 2003. See also E. O. Wilson and B. Holldobler, "The Rise of the Ants: A Phylogenetic and Ecological Explanation," *Proceedings of the National Academy of Sciences* 102 (2005): 7,411–14; and E. O. Wilson and B. Holldobler, "Eusociality: Origin and Consequences," *Proceedings of the National Academy of Sciences* 102 (2005): 13,367–71.

13. Wilson, *Sociobiology*, 554.

14. R. L. Trivers, "Parent-Offspring Conflict," *American Zoologist* 14 (1974): 249–65.

15. Wilson, *Sociobiology*, 555.

16. R. Dawkins, *The Selfish Gene*, 2nd ed. (Oxford: Oxford University Press, 1989), 328.

17. J. L. Brown, *Helping and Communal Breeding in Birds* (Princeton, N.J.: Princeton University Press, 1987); P. Stacey and W. Koenig, eds.,

Cooperative Breeding in Birds: Long-Term Studies of Ecology and Behavior (Cambridge: Cambridge University Press, 1990).

18. J. L. Brown, E. Brown, S. D. Brown, and D. D. Dow, "Helpers: Effects of Experimental Removal on Reproductive Success," *Science* 215 (1982): 421–22.

19. A queen has two sets of chromosomes and hence two sets of genes. Both male and female offspring each receive a single set of these genes, so the queen shares copies of half her genes with each offspring, regardless of sex.

20. Fisher, *Genetical Theory of Natural Selection*.

21. R. Trivers and H. Hare, "Haplodiploidy and the Evolution of the Social Insects," *Science* 191 (1976): 249–63.

22. Trivers and Hare remind their readers of something that often gets lost in reading Hamilton's original models; namely that a 3:1 sex ratio is not only the expected outcome if workers are in control, but that such a sex ratio is also critical for the very altruism that in many ways defines the Hymenoptera. If workers are related to their sisters by an r of 0.75, and to their brothers by an r of 0.25, then when the sex ratio in a colony is 1:1, the average relatedness of workers to their siblings is 0.5. Under such a scenario a worker is no more related to her siblings ($r = 0.5$) than to her offspring ($r = 0.5$), and hence there is no selective pressure on workers to help raise their siblings at the expense of their own offspring. If, however, there are three females to every male in the hive, then a worker who invests in helping the queen produce *lots* of new sister workers ($r = 0.75$) will be favored over a worker who attempts to produce offspring of her own.

23. Hamilton, *Narrow Roads*, 1:2.

CHAPTER EIGHT
KEEPERS OF THE FLAME

1. When Alexander taught his first course in animal behavior, Emlen was his teaching assistant.

2. Author's interview with Stephen Emlen, Cornell University, Ithaca, N.Y., January 31, 2003.

3. E-mail message from Stephen Emlen, February 17, 2003.

4. Approximately half of all nests have helpers at one time or another. S. T. Emlen and P. H. Wrege, "A Test of Alternative Hypotheses for Helping Behavior in White-fronted Bee-eaters," *Behavioral Ecology and Sociobiology* 25 (1989): 705. The bee-eater is one of the few bird

species that is both a cooperative and colonial breeder. S. T. Emlen and N. Demong, "Bee-eaters: An Alternative Route to Cooperative Breeding," paper presented at the Symposium on Altruism in Birds, 27th International Ornithological Congress, West Berlin, 1978.

5. Little did he know that he would be making such visits each spring for more than the next twenty years.

6. Emlen interview, January 31, 2003.

7. Emlen's work has also uncovered a conflict of interest in bee-eaters. Like the young of many communally breeding species, young male white-fronted bee-eaters (*Merops bullockoides*) sometimes forgo the opportunity to breed on their own. These males become helpers, remaining on their natal territory and aiding their blood relatives, usually their parents, in raising young. When breeding opportunities for young males are rare, no conflict exists between young male helpers and their parents: it is in the best interest of both parties for the male to remain at home and assist his parents. When breeding opportunities away from the natal nest become more readily available, young males are favored to breed on their own, but it may be in the best interest of such a male's parents for him to remain at home and help raise his siblings. S. T. Emlen and P. H. Wrege, "Parent-Offspring Conflict and the Recruitment of Helpers among Bee-eaters," *Nature* 356 (1992): 333.

8. S. T. Emlen, "An Evolutionary Theory of the Family," *Proceedings of the National Academy of Sciences* 92 (1995): 8,092–99.

9. The CGSS surveyed 13,495 households.

10. J. Davis and M. Daly, "Evolutionary theory and the human family," *Quarterly Review of Biology* 72 (1997): 407–35.

11. Emlen, "An Evolutionary Theory of the Family."

12. S. Pruett-Jones and M. Lewis, "Sex Ratio and Habitat Limitation Promote Delayed Dispersal in Superb Fairy Wrens," *Nature* 348 (1990): 541–42.

13. Indeed, in some cases (males, age 15–24), married individuals living away from home were more likely to remain in contact with their parents than were singles.

14. Instead, Davis and Daly suggest that that human parents act as postreproductive helpers to their own offspring. This help to children and grandchildren may have selected for strong family bonds that do not easily dissolve when offspring get married.

15. Emlen, "An Evolutionary Theory of the Family."

16. See J. L. Brown, "Alternate Routes to Sociality in Jays—with a Theory for the Evolution of Altruism and Communal Breeding," *American*

Zoology 14 (1974): 63–80. Not only are the offspring that remain on high-quality territories receiving a benefit, but their parents are as well, since they pass down the best-quality territories to their blood kin.

17. P. B. Stacey and J. D. Ligon, "Territory Quality and Dispersal Options in the Acorn Woodpecker, and a Challenge to the Habitat-Saturation Model of Cooperative Breeding," *American Naturalist* 130 (1987): 654–76.

18. Moreover, male birds who served as helpers had a relatively high probability of entering the breeding population, often breeding in turn on their natal territory. In another cooperatively breeding bird species, the Florida scrub jay, multigenerational dynasties are created on high-quality home areas via territory "budding." Budding occurs when aggression on the part of the territory holder and his family produces a new, expanded area that includes space for newly developing breeders.

19. L. White, "Coresidence and Leaving Home: Young Adults and Their Parents," *Annual Review of Sociology* 20 (1994): 81–102. Reviewed by Davis and Daly and used to test the hypothesis.

20. Using the CGSS data, Davis and Daly addressed the more detailed question of whether contact with kin is not only more likely, but more frequent, as a function of wealth. Using letter, phone, or face-to-face conversations as a measure of contact, they found that for most age and sex cohorts, wealthier individuals did keep in touch with relatives more often than did lower-income individuals. Within the subset of individuals who did make contact with family members, however, poorer individuals, made *more* contact with family members than their richer cohorts, suggesting that the original differences resulted from many poorer individuals making no contact at all with family members. See also L. White and A. Reidmann, "Ties among Adult Siblings," *Social Forces* 71 (1992): 85–102; M. Taylor, "Receipt of Support from Family among Black Americans: Demographic and Familial differences," *Journal of Marriage and the Family* 48 (1986): 67–77; D. Eggebeen and D. Hogan, "Giving between Generations in American Families," *Human Nature* 1 (1990): 211–32.

21. Emlen, "An Evolutionary Theory of the Family."

22. The extreme version of this hypothesis predicts that, under certain conditions, natural selection might even favor a stepparent's killing its offspring. If the stepparent is a male, patricide may result in his mate's quickly returning to an estrous state.

23. There is growing evidence that avian stepparents also destroy clutches containing potential stepchildren.

24. G. Hausfater and S. Blaffer-Hrdy, *Infanticide* (New York: Aldine, 1984).

25. J. Bray, "Children's Development during Early Remarriage," in *Impact of Divorce, Single Parenting and Stepparenting on Children*, ed. E. Hetherton and J. Arasteh (Hillsdale, N.J.: Lawrence Earlbuam, 1988), 279–98; L. White, "Stepfamilies over the Life Course: Social Support," in *Stepfamilies: Who Benefits? Who Does Not?*, ed. A. Booth and J. Dunn (Hillsdale, N.J.: Lawrence Earlbaum, 1994), 109–37.

26. M. Daly and M. Wilson, *Homicide* (New York: Aldine de Gruyter, 1988); M. Daly and M. Wilson, *The Truth about Cinderella: A Darwinian View of Parental Love* (New Haven, Conn.: Yale University Press, 1999). Daly and Wilson are well aware that confounding factors may sometimes make it look as if stepparenting is associated with increased levels of violence toward stepchildren, when it is not. They have systematically addressed such concerns. For example, they have demonstrated that the connection between stepparent violence cannot be ascribed to stepfamilies being likely to live in poverty. In addition, critics have claimed that violence in stepfamilies may be reported more often than in families with no stepparents, creating a reporting bias. Daly and Wilson reasoned that if this was correct, such a bias would decrease with the severity of the violent act. In the extreme case—murder of a child—surely this bias would disappear. Yet, when they narrowed down the child abuse cases from 87,789 to the 279 that involved murder, they found that the risk ratio for stepchildren actually increased. *The Truth about Cinderella*, 28.

27. Daly and Wilson, *The Truth about Cinderella*, 7.

28. Daly and Wilson, *Homicide*, and *The Truth about Cinderella*; D. Kaplun and R. Reich, "The Murdered Child and His Killers," *American Journal of Psychiatry* 133 (1976): 809–13; M. Daly and M. Wilson, "Some Differential Attributes of Lethal Assaults on Small Children by Stepfathers versus Genetic Fathers," *Ethology and Sociobiology* 15 (1994): 207–17. But see H. Temrin, S. Buchmayer, and M. Enquist, "Stepparents and Infanticide: New Data Contradict Evolutionary Predictions," *Proceedings of the Royal Society of London* 267 (2000): 943–45.

29. Daly and Wilson, *The Truth about Cinderella*, 50. Given the high rates of child abuse, it is hardly surprising that children from stepfamilies leave home significantly earlier than children from biologically intact families. The comparison between step- and biological families is not restricted to parent-offspring interactions, as stepsiblings also have a lower r (0 or 0.25) than biological siblings ($r = 0.5$). In line with evolutionary thinking, the data on sibling-sibling aggression suggests

that children behave less kindly toward their half-sibs than their full-sibs. N. Zill, "Behavior, Achievement and Health Problems among Children in Stepfamiles: Findings from a National Survey on Child Health," in *Impact of Divorce, Single Parenting and Stepparenting on Children*, 325–68; W. Aquilino, "Family Structure and Home Leaving: A Further Specification of the Relationship," *Journal of Marriage and the Family* 53 (1991): 999–1,010.

30. From an e-mail exchange with Margo Wilson, December 9, 2004.

31. Emlen interview, January 31, 2003.

32. Sherman's dissertation work was done at the University of Michigan. He first learned of the ground squirrel system from Martin Morton.

33. P. W. Sherman, "Nepotism and the Evolution of Alarm Calls," *Science* 197 (1977): 1,246–53. This was the first of a series of such papers that Sherman would publish, including: "The Limits of Ground Squirrel Nepotism," in *Sociobiology: Beyond Nature/Nurture?*, ed. G. Barlow and J. Silverberg (Boulder, Colo.: Westview, 1980), 505–44; "The Meaning of Nepotism," *American Naturalist* 116 (1980): 604–6; "Kinship, Demography, and Belding's Ground Squirrel Nepotism," *Behavioral Ecology and Sociobiology* 8 (1981): 251–59; "Alarm Calls of Belding's Ground Squirrels to Aerial Predators: Nepotism or Self-preservation?" *Behavioral Ecology Sociobiology* 17 (1985): 313–23; and P. W. Sherman and W. Holmes, "Kin Recognition: Issues and Evidence," *Fortschritte der Zoologie* 31 (1985): 437–60.

34. Author's interview with Paul Sherman, Cornell University, Ithaca, N.Y., January 31, 2003.

35. Sherman examined five other possible benefits that terrestrial alarm callers—particularly females that remained on their natal territory—might receive. Alarm calls might, for example, divert the predator's attention, discourage the predator from attacking, or reduce the chance of future attack by the same predator. None of these alternative hypotheses were supported by the data. Alarm callers were not somehow safer by way of calling, nor did they deter predators from attacking. Instead, blood relatedness, in conjunction with the demographics mentioned above, were the keys to understanding terrestrial alarm calls in Belding's ground squirrels.

36. Sherman interview, January 31, 2003.

37. Sherman, "Alarm Calls of Belding's Ground Squirrels to Aerial Predators," 313.

38. Sherman interview, January 31, 2003.

39. Termites are part of the insect order Isoptera.

40. P. W. Sherman, J. Jarvis, and R. Alexander, eds. *The Biology of the Naked Mole-Rat* (Princeton, N.J.: Princeton University Press, 1991), ix.

41. Sherman interview, January 31, 2003.

42. Compared with other species, naked mole-rats lead extraordinarily long lives. P. Sherman and J. Jarvis, "Extraordinary Life Spans of Naked Mole-rats (*Heterocephalus glaber*)," *Journal of the Zoological Society of London* 258 (2002): 307–11.

43. E. A. Lacey, R. D. Alexander, S. H. Braude, P. W. Sherman, and J. Jarvis, "An Ethogram for the Naked Mole-rat: Non-vocal Behaviors," in *The Biology of the Naked Mole-Rat*, 209–42. E. A. Lacey and P. W. Sherman, "Social Organization of Naked Mole-rats: Evidence for Divisions of Labor," in *The Biology of the Naked Mole-Rat*, 275–336. H. K. Reeve and P. W. Sherman, "Intracolonial Aggression and Nepotism by the Breeding Female Naked Mole-rat," in *The Biology of the Naked Mole-Rat*, 337–57.

44. J.U.M. Jarvis, "Eusociality in a Mammal: Cooperative Breeding in Naked Mole-rat Colonies," *Science* 212 (1981): 571–73.

45. H. K. Reeve, D. F. Westneat, W. A. Noon, P. W. Sherman, and C. F. Aquadro, "DNA 'Fingerprinting' Reveals High Levels of Inbreeding in Colonies of the Eusocial Naked Mole-rat," *Proceedings of the National Acadamy of Sciences* 87 (1990): 2,496–2,500.

46. Reeve et al., "DNA 'Fingerprinting,'" 2,499. Why the deleterious effects typically associated with inbreeding (for example, an increase in diseases caused by recessive alleles) are absent in naked mole-rats remains a mystery.

47. H. K. Reeve, "Queen Activation of Lazy Workers in Colonies of the Eusocial Naked Mole-rat," *Nature* 358 (1992): 147–49.

48. When a queen is removed from a colony, it is the workers that are least related to her that are most likely to decrease their workload.

49. D. Fletcher and C. Michener, *Kin Recognition in Animals* (New York: Wiley, 1987); P. G. Hepper, *Kin Recognition* (Cambridge: Cambridge University Press, 1991); D. W. Pfennig and P. W. Sherman, "Kin Recognition," *Scientific American*, June 1995, 98–103; P. Hepper and J. Cleland, "Developmental Aspects of Kin Recognition," *Genetica* 104 (1998): 199–205.

50. B. Waldman, "Mechanisms of Kin Recognition," *Journal of Theoretical Biology* 128 (1987): 159–85; H. K. Reeve, "The Evolution of Conspecific Acceptance Thresholds," *American Naturalist* 133 (1989): 407–35.

51. The internal template of an individual may be generated by either some sort of genetically based rule or via learning. E. Boyse,

G. Beauchamp, K. Yamazaki, and J. Bard, "Genetic Components of Kin Recognition in Mammals," in *Kin Recognition*, 162–219; R. Crozier, "Genetic Aspects of Kin Recognition: Concepts, Models and Synthesis," in *Kin Recognition in Animals*, 55–74; S. Robinson and W. Smotherman, "Fetal Learning: Implications for the Development of Kin," in *Kin Recognition*, 308–34; R. D. Alexander, *Darwinism and Human Affairs* (Seattle: University of Washington Press, 1979); R. D. Alexander, "Social Learning and Kin Recognition," Ethology and Sociobiology 12 (1991): 387–99.

52. Reeve, "Evolution of Conspecific Acceptance Thresholds." Reeve's "conspecific threshold model" is the most detailed and predictive of all the internal template models. It uses sophisticated mathematics to determine just how similar individual 2 must to be to individual 1's template before it is considered kin. Reeve found that such "similarity values" are a function of (1) The relative frequency at which individuals interact with kin and nonkin. If the majority of an individual's interactions are with blood relatives, there is little selection pressure to generate strict rules as to who are and who are not kin. (2) The fitness consequences of rejecting blood kin by mistake, or accepting an individual as kin when it in fact is not a blood relative. When the cost of making a mistake is high, natural selection favors individuals who categorize only those who closely match their own internal template as kin.

53. D. W. Pfennig, H. K. Reeve, and P. W. Sherman, "Kin Recognition and Cannibalism in Spadefoot Toads,"*Animal Behavior* 46 (1993): 87–94; M. Elgar and B. Crespi, *Cannibalism: Ecology and Evolution among Diverse Taxa* (Oxford: Oxford University Press, 1992).

54. W. Holmes and P. W. Sherman, "Kin Recognition in Animals," *American Scientist* 71 (1983): 46–55; P. W. Sherman and W. Holmes, "Kin Recognition: Issues and Evidence," *Fortschritte Zoologie* 31 (1985): 437–60; A. Blaustein, "Kin Recognition Mechanisms: Phenotype Matching or Recognition Alleles," *American Naturalist* 121 (1983): 749–54; A. Blaustein, M. Bekoff, J. Byers, and T. Daniels, "Kin Recognition in Vertebrates: What Do We Really Know about Adaptive Values?" *Animal Behavior* 41 (1991): 1,079–83.

55. M. D. Beecher, I. Beecher, and S. Lumpkin, "Parent-Offspring Recognition in Bank Swallows: I. Natural History," *Animal Behavior* 29 (1981): 86–94; M. D. Beecher, I. Beecher, and S. Hahn, "Parent-offspring recognition in bank swallows: II. Development and acoustic basis," *Animal Behavior* 29 (1981): 95–101; M. D. Beecher, B. Medvin, P. Stoddard, and P. Loesch, "Acoustic Adaptations for Parent-Offspring Recognition in Swallows," *Experimental Biology* 45 (1986): 179–93.

CHAPTER NINE
CURATOR OF MATHEMATICAL MODELS

1. Hamilton, *Narrow Roads*, 1:486.

2. A faculty member at Michigan's Institute for Public Policy Studies (now called the School of Public Policy).

3. Axelrod and Hamilton were not the first researchers to suggest that the prisoner's dilemma game could be used to address the evolution of cooperation and altruism. In a 1971 paper entitled "The Evolution of Reciprocal Altruism," Robert Trivers hypothesized that natural selection would favor genes for altruistic behavior *if* individuals were altruistic only toward those who had helped them in the past—that is, if altruism was reciprocated. Trivers then suggested that the prisoner's dilemma game was a useful tool for examining the evolution of such reciprocal altruism. R. L. Trivers, "The Evolution of Reciprocal Altruism," *Quarterly Review of Biology* 46 (1971): 189–226.

4. Axelrod had a long-standing interest in evolutionary biology. In 1963, as an undergraduate at the University of Chicago, he had done an internship at Rashevsky's Institute for Mathematical Biology (from e-mail exchange with Robert Axelrod, September 2, 2003).

5. Hamilton, *Narrow Roads*, 2:119.

6. Hamilton had some experience with game theory, reciprocity, and the prisoner's dilemma. W. D. Hamilton, "Extraordinary Sex Ratios," *Science* 156 (1967): 477–87; W. D. Hamilton, "Altruism and Related Phenomena, Mainly in Social Insects," *Annual Review of Ecological Systems* 3 (1972): 192–232; W. D. Hamilton, "Selection of Selfish and Altruistic Behaviour in Some Extreme Models," in *Man and Beast: Comparative Social Behavior*, ed. J. F. Eisenberg and W. S. Dillon (Washington, D.C.: Smithsonian Institution Press, 1971), 57–91.

7. Hamilton, *Narrow Roads*, 2:120.

8. W. Poundstone, *Prisoner's Dilemma: Jon Von Neuman, Game Theory and the Puzzle of the Bomb* (New York: Doubleday, 1992).

9. More precisely, they were searching for the evolutionarily stable strategy (ESS) to this game. An ESS is a strategy that at high frequencies cannot be "invaded" and replaced by any other strategy.

10. In fact, Axelrod and Hamilton ran a second computer tournament with sixty-two strategies. Results were similar to those in the initial tournament.

11. Submitted by Anatol Rapoport of the Institute of Advanced Study in Vienna.

12. It is also the case that a strategy called "always defect" also does well in the iterated prisoner's dilemma game.

13. Since the original models of *tft* appeared, dozens of variants of this strategy have been examined, but most share the essential characteristics; see L. A. Dugatkin, *Cooperation among Animals: An Evolutionary Perspective* (New York: Oxford University Press, 1997) for more on this.

14. Hamilton, *Narrow Roads*, 2:124.

15. http://www.royalsoc.ac.uk/funding/ResProf/CondTenRP1A.pdf, condition 5.

16. Hamilton, *Narrow Roads*, 2:303.

17. Ibid., 304, 307–8.

18. Richard Dawkins in a forward to Hamilton, *Narrow Roads*, 2:xvii.

19. Hamilton, *Narrow Roads*, 2:100.

20. The second volume of the Hamilton's collected papers is subtitled "Evolution of Sex." But the evolution of sex was not the only thing on his mind. He developed and tested new hypotheses concerning the wondrous colors of autumn foliage—which Hamilton believed were an elaborate signalling system by which trees advertised their health to potentially dangerous insects—and whether clouds were transmission vectors for algae and marine microbes. W. D. Hamilton and S. Brown, "Autumn Tree Colors as a Handicap Signal," *Proceedings of the Royal Society of London* 268 (2001): 1,489–93; W. D. Hamilton and T. M. Lenton, "Spora and Gaia: How Microbes Fly with Their Clouds," *Ethology Ecology & Evolution* 10 (1998): 1–16.

21. Hamilton was still able to recognize friends and family.

22. E. Hooper, *The River: A Journey Back to the Source of HIV and AIDS* (New York: Little, Brown, 1999).

23. Alan Grafen, *Guardian*, March 9, 2000.

24. *London Times*, March 9, 2000.

25. Natalie Angier, *New York Times*, March 10, 2000.

26. Richard Dawkins, *Independent* (London), March 10, 2000.

27. Hamilton, *Narrow Roads*, 1:93.

▲ *Index* ▲

CPSIA information can be obtained
at www.ICGtesting.com
Printed in the USA
JSHW081450071022
31411JS00004B/6

9 780691 242132